13th International Conference on Environment Pollution and Prevention (ICEPP 2025)

AA004287

E3S Web of Conferences Volume 690 (2026)

Ho Chi Minh City, Vietnam
6-8 November 2025

Editor:

R.J. Haynes

ISBN: 979-8-3313-3270-9

Printed from e-media with permission by:

Curran Associates, Inc.
57 Morehouse Lane
Red Hook, NY 12571

Some format issues inherent in the e-media version may also appear in this print version.

This work is licensed under a Creative Commons Attribution 4.0 International License. License details: http://creativecommons.org/licenses/by/4.0/.

No changes have been made to the content of these proceedings. There may be changes to pagination and minor adjustments for aesthetics.

Printed with permission by Curran Associates, Inc. (2026)

For additional information, please contact EDP Sciences – Web of Conferences at the address below.

EDP Sciences – Web of Conferences
17, Avenue du Hoggar
Parc d'Activité de Courtabœuf
BP 112
F-91944 Les Ulis Cedex A
France

Phone: +33 (0) 1 69 18 75 75
Fax: +33 (0) 1 69 28 84 91

contact-edps@webofconferences.org

Additional copies of this publication are available from:

Curran Associates, Inc.
57 Morehouse Lane
Red Hook, NY 12571 USA
Phone: 845-758-0400
Fax: 845-758-2633
Email: curran@proceedings.com
Web: www.proceedings.com

TABLE OF CONTENTS

GIS-Based Mapping and Profiling of Forest Cover and Freshwater Resources for Sustainable and Climate-Resilient Resource Management in the Island Barangays of Surigao City, Philippines 1
Dela Cerna Monalee, Odtojan Medielyn, Degamon Louella, Cuadrado Jerry, Bertulfo Roselle, Hugo Rosalia, Eviota Manny, Ecle Kier, Adlaon Mauricio, Mosa Carmelin

Afforestation of Tropical Watersheds to Minimize Flood Risk in the Parish of Valle Hermoso, Ecuador 11
Jiménez-Pozo Patricio, Argotti-Valencia Eduardo, Suasnavas Darwin, Bautista Hugo

Review of Strategies to Control Air Pollution Due to PFAS 17
Rajurkar Prashant

Vacuum–Suction Technology Integrated with Water, Gas and Solid Waste Treatment: A Comprehensive Solution for Urban and Industrial Environments in Vietnam 27
Luong Nguyen Duc, Ho Huynh Kim, Hieu Huynh Thuan, Y Nguyen Nhu

Integrating Indigenous Governance into Nature-Based Solutions for Climate and Biodiversity Resilience 36
Nur Rismawati, Meilani Retno, Dwi Novemyanto Alfin

Integrating Waste Valorization and Green Energy: Biodiesel Production Using Microwave-Derived Chitosan Catalyst from Raw Tilapia (Oreochromis Niloticus) Scales 46
Panaligan Tristan Roy, Alam Andrea Marie, Berroya Jean Margaret, Torres Princess Lynelle Carysse

How to Reduce Environmental Burden Caused by Lignocellulosic Waste: A Solution from the Pharmaceutical Industry with Green Solvents 52
Yu Lijiao, Xia Yu, Zhu Ruiliang, Yao Shun

Fabrication of Eco-Friendly Pineapple Leaf Fiber-Based Vegan Leather for Environmental Sustainability 60
Srisang Siriwan, Kunyuan Jumpon, Hutangkoon Thunyanat, Eiangmee Orranat, Maikaew Jatuphat, Sokjabok Siwakon, Srisang Naruebodee

Performance Compatibility Assessment of Selected Motorcycles Using Nipa Bioethanol as Alternative Fuel Blend 76
Mateo Nathaniel Ericson, Calderon Aldrin, Agrupis Shirley, Acidera Kendrick Cyrus, Manzano Loreli Faye

Characterization and Performance of Pineapple Stem Cationic Starch as an Eco-Friendly Flocculant for Phosphate Removal from Wastewater 88
Erasga Eden S., Llames Angela Mae A., Sangab Kim Z., Yap Angelica Nala T.

Assessing the Impact of Battery Electric Vehicles (BEVs) Adoption on Air Quality and Public Health in Indonesia 96
Ghaida Azka, Ningsih Almira Solehah, Dewati Regyna Elrica, Muharrami Revia, Permadi Didin Agustian, Goyal Prachi, Kusuma Ririn Radiawati

Investigation of Heavy Metals in Indoor Dusts of Households in an Indubitable Range from a Landfill Located in South-Western India 105
Lal Naveen S., Aravindakumar C.T, Aravind Usha K.

The Effects of Indoor Air Pollution from Solid Fuel Burning on Neurological and Neuropsychiatric Diseases - A Review of Epidemiological Evidence 115
Yang Jianing, Fan Minjia, Xu Linglong

Recycling Library as Fulfillment of the Right to Education and a Healthy Sustainable Environment 134
Novemyanto Alfin Dwi, Nur Rismawati, Meilani Retno

Rethinking Carbon Pricing: Balancing Market Incentives and Environmental Protection Through BUMICX 143
Meilani Retno

IoT-Based Multimodal Borewell Monitoring 152
Bhatt Tanmay, Reddy Jagan M., Pragada Tarun, Chaudhari Sachin, Ramesh Kalpana

CrackVision: A Bayesian CNN–Vision Transformer Bagging Ensemble for Enhancing Multiclass Concrete Crack Severity Classification 163
Mandap Jomer R., Sanchez Daniel John Henrick D., Ontanillas Steven Mark B., Magloyuan Neil P.

Author Index

ICEPP 2025

2025 13th International Conference on
ENVIRONMENT POLLUTION AND PREVENTION

Ho Chi Minh City, Vietnam November 6-8, 2025

Co-sponsored by

ICED

BK TP.HCM

Supported by

THE UNIVERSITY OF QUEENSLAND AUSTRALIA

UNIVERSITAS GALATIENSIS

REXDAN

COURAGE TO KNOW

PCCCC-II.CHD

DELTECH

UTM School of Mechanical Engineering UNIVERSITI TEKNOLOGI MALAYSIA

Statement of Peer review

In submitting conference proceedings to E3S Web of Conferences, I certify to the Publisher that I adhere to the **Policy on Publishing Integrity** of the journal in order to safeguard good scientific practice in publishing.

1. All articles have been subjected to peer review administered by the proceedings editors.
2. Reviews have been conducted by expert referees, who have been requested to provide unbiased and constructive comments aimed, whenever possible, at improving the work.
3. Proceedings editors have taken all reasonable steps to ensure the quality of the materials they publish and their decision to accept or reject a paper for publication has been based only on the merits of the work and the relevance to the journal.

Title, date and place of the conference

2025 13th International Conference on Environment Pollution and Prevention (ICEPP 2025)
Ho Chi Minh City, Vietnam, November 6-8, 2025

Proceedings editor(s): R. J. (Dick) Haynes

Date and editor's signature 2026.1.8

https://doi.org/10.1051/e3sconf/202669000001

Preface

2025 13th International Conference on Environment Pollution and Prevention (ICEPP 2025) was successfully convened from November 6-8, 2025, in Ho Chi Minh City, Vietnam. As an annual event launched in 2013, ICEPP continues to serve as a key forum for leading academics and industry professionals dedicated to addressing environmental challenges. This year's program provided a dynamic platform for presenters to share past achievements, renew professional connections, and explore emerging research directions together. We trust that all participants found the conference both intellectually stimulating and engaging.

ICEPP 2025 featured four distinguished keynote speakers, two invited speakers and was organized into two in-person sessions and one virtual session. Thematic discussions spanned Environmental Pollution Monitoring, Control, and Health Impact Assessment, Sustainable Resource Management and Ecological Environment Protection, as well as Sustainable Environmental Technologies and Pollution Control. The conference brought together delegates from diverse regions including Australia, China, Vietnam, Indonesia, India, Malaysia, Philippines, Korea, Kazakhstan, Ethiopia and beyond, fostering meaningful dialogue on contemporary innovations, pressing issues, and practical solutions within the field.

This proceedings volume comprises a curated selection of accepted papers, each having undergone a rigorous peer-review process to ensure the inclusion of high-quality and original research that reflects the core outcomes of the conference.

We extend our sincere gratitude to all participants, sponsors, and supporters whose dedication and enthusiasm were instrumental in making ICEPP 2025 a notable success. It is our hope that this collection will offer valuable insights and inspire further scholarly and practical advances among its readers.

Proceedings Editor
R. J. (Dick) Haynes
The University of Queensland, Australia

© The Authors, published by EDP Sciences. This is an open access article distributed under the terms of the Creative Commons Attribution License 4.0 (https://creativecommons.org/licenses/by/4.0/).

Committees

Honorary Chairs

Mai Thanh Phong, President/ Rector of Ho Chi Minh City University of Technology (HCMUT), VNU-HCM, Vietnam
Huynh Ky Phuong Ha, Dean, Faculty of Environment and Natural Resources (FENR – HCMUT), Vice-Rector of HCMUT, VNU-HCM, Vietnam

Conference Chairs

Sunil Herat, Griffith University, Australia
Nguyen Hong Quan, Nguyen Hong Quan, Institute for Circular Economy Development (ICED), VNU-HCM, Vietnam
Vo Le Phu, FENR – HCMUT, VNU-HCM, Vietnam

Conference Co-Chairs

Xingbin Sun, Northeast Forestry University, China
Dennis Y.C. Leung, University of Hong Kong, Hong Kong, China

Local Organizing Committee Chairs

Nguyen Nhat Huy, FENR – HCMUT, Vietnam
Vo Le Phu, FERN – HCMUT, Vietnam

Conference Program Chairs

Rouzbeh Abbassi, Macquarie University, Sydney, Australia
Sue Charlesworth, Coventry University, United Kingdom
Roland Kallenborn, Norwegian University of Life Sciences, Norway

Conference Steering Chair

R. J. (Dick) Haynes, The University of Queensland, Australia

Conference Publicity Chairs

Lai Quoc Dat, External Affairs Office – HCMUT, Vietnam
Pham Tan Thi, Science and Technology Office – HCMUT, Vietnam
Tran Minh Quang, External Affairs Office – HCMUT, Vietnam

Technical Committee

Eiji Haramoto, University of Yamanashi, Japan

Ghani Albaali, Princess Sumaya University for Technology PSUT, Jordan
Angela T Ragusa, Charles Sturt University, NSW Australia
Chun-Hung(Jeb) Lee, Dean, Center for Sustainable Development, National DongHwa University, Taiwan
Nilofar Asim, University Kebangsaan Malaysia (UKM), Malaysia.
Gargh Shankar, Environmental Disasters Research Institute, India
Seongjin Hong, Chungnam National University, Korea
Md Kamrul Hasan, Zhejiang University, China
Saikat Sinha Ray, Ulsan National Institute of Science and Technology, South Korea
Fu-Ming Chang, Da-Yeh University, Taiwan, China
Dinesh K. Yadav, ICAR-National Academy of Agricultural Research Management, India
Meisam Ahmadi Ghadikolaei, The Hong Kong University of Science and Technology, Hong Kong, China
Md. Mizanur Rahman, University Technology Malaysia, Malaysia
Mylene G. Dinglasan, Philippine Christian University, Philippine
Linhui Zeng, Hangzhou XiHu District City Management and Security Service Center, China
Hongwen Sun, Nankai University, China
Qing Zhu, Purdue University, USA
Shahjalal Khandaker, Dhaka University of Engineering & Technology, Gazipur, Bangladesh
Zhaoyang Liu, Qatar Environment and Energy Research Institute, Hamad Bin Khalifa University (HBKU), Qatar Foundation
Xin Hu, Beijing University of Technology, China

GIS-based mapping and profiling of forest cover and freshwater resources for sustainable and climate-resilient resource management in the island barangays of Surigao City, Philippines

Monalee dela Cerna[1*], *Medielyn* Odtojan[2], *Louella* Degamon[3], *Jerry* Cuadrado[2], *Roselle* Bertulfo[2], *Rosalia* Hugo[2], *Manny* Eviota[3], *Kier* Ecle[3], *Mauricio* Adlaon[3], and *Carmelin* Mosa[3]

[1]College of Computing and Information Sciences, Surigao del Norte State University, Surigao City, Philippines
[2]College of Arts and Sciences, Surigao del Norte State University, Surigao City, Philippines
[3]College of Teacher Education, Surigao del Norte State University, Surigao City, Philippines

Abstract. The increasing environmental challenges in the island barangays of Surigao City, including deforestation, freshwater scarcity, and climate change impacts, necessitate a systematic approach to resource management. This study utilized Geographic Information Systems (GIS) to map and analyze the forest cover and freshwater resources of Nonoc, Hikdop, and Bayagnan Islands. Data was collected through satellite imagery, field surveys, and Digital Elevation Models (DEM), and Normalized Difference Vegetation Index (NDVI) to assess vegetation distribution, water sources, topography, and terrain characteristics. Findings reveal that most barangays exhibit tertiary growth forests, with freshwater primarily sourced from natural springs and deep wells. GIS-based analysis of elevation, slope, and aspect provided critical insights into potential environmental risks such as flooding, erosion, and land degradation. The study highlights the need for integrated GIS monitoring systems, sustainable water resource management, reforestation initiatives, and disaster risk mitigation strategies. These findings serve as a valuable tool for policymakers, conservationists, and local communities in fostering evidence-based environmental planning and sustainable development.

1 Introduction

Water security has emerged as a critical global concern, encompassing sustainable management, accessibility, and quality of water resources in the face of increasing demand, climate variability, and environmental degradation. Rapid population growth, urbanization, and industrial expansion are intensifying pressure on water availability and distribution [1]. Climate change further compounds these challenges by altering precipitation patterns, increasing the frequency of extreme weather events, and disrupting hydrological cycles [2]. Additionally, pollution, contamination, and inadequate sanitation systems continue to degrade water quality, posing serious threats to human health and ecological balance [3].

* Corresponding author: mdelacerna@ssct.edu.ph

© The Authors, published by EDP Sciences. This is an open access article distributed under the terms of the Creative Commons Attribution License 4.0 (https://creativecommons.org/licenses/by/4.0/).

In response, United Nations established Sustainable Development Goal 6 (SDG 6), which seeks to ensure universal access to clean water and sanitation by 2030. However, despite global efforts, nearly 40% of the global world's population still lacks access to safe drinking water and reliable sanitation services [4]. At the national level, the Philippine National Integrated Basic Research Agenda (NIBRA) prioritizes water security through the TUBIG Agenda (Tubig Ay Buhayin At Ingatan), which anticipates a severe water shortage by 2040 due to climate change and population growth. This agenda underscores the importance of developing new technologies and research-based strategies for water detection, management, and conservation. Ongoing initiatives focus on enhancing improving water quality monitoring, improving resource allocation, and formulating policy recommendations particularly for coastal and island communities vulnerable to saltwater intrusion and freshwater depletion [5].

The integration of Geographic Information System (GIS) in environmental research has transformed natural resource management by providing precise, data-driven insights. GIS plays a pivotal role in mapping and profiling forest cover and freshwater resources enabling an interdisciplinary approach that merges geospatial analysis, environmental science, and resource management. Understanding the extent and characteristics of forested areas is vital for biodiversity conservation, climate change mitigation, and land-use planning [6]. GIS-based mapping allows for the accurate delineation of forests, enabling policymakers to monitor deforestation trends, identify biodiversity hotspots, and track land-use changes over time. Rivers, lakes, and aquifers are essential for ecosystems, livelihoods, and industries, and GIS supports spatial analyses of water distribution while identifying threats from pollution, over-extraction, and climate change [7]. Forests and freshwater ecosystems are deeply connected—vegetation regulates hydrological cycles and helps maintain water quality [8]. GIS-based analyses therefore enhance understanding of these interdependencies, leading to more holistic conservation strategies.

Despite these technological advancements, challenges persist in ensuring data accuracy, overcoming remote sensing limitations, and fostering interdisciplinary collaboration. In the island barangays of Surigao City, the absence of an integrated GIS-based mapping system for forest cover and freshwater resources limits effective conservation and sustainable resource management. Existing data are often outdated, fragmented, or incomplete, hindering policymakers' ability to design targeted environmental interventions. Without accurate geospatial information, critical issues such as deforestation, water scarcity, and ecological degradation remain poorly understood and inadequately addressed.

The island barangays of Surigao City possess a unique ecological landscape where forest cover and freshwater resources are fundamental to environmental sustainability and human survival. However, rapid urbanization, climate change, and various anthropogenic activities have disrupted the delicate balance of these resources, threatening biodiversity, water availability, and ecological stability. Without a systematic, data-driven approach to monitoring these natural assets, the risk of environmental degradation and water scarcity continues to escalate, jeopardizing both ecosystems and local communities.

Recognizing the urgency of theses challenges, this study aims to harness the power of GIS to conduct a comprehensive spatial analysis of the distribution, interrelationships, and condition of forested areas and freshwater resources in the island barangays of Surigao City. By integrating advanced GIS technologies with field-based data, this research seeks to generate accurate geographic information that can inform decision-making, guide resource conservation efforts, and support the development of evidence-based environmental policies.

 https://doi.org/10.1051/e3sconf/202669001001

2 Methods

This study employed a structured GIS and spatial analysis framework to map forest cover and freshwater resources in the island barangays of Surigao City. Satellite imagery, aerial photographs, and field and GPS-referenced surveys were used to generate thematic layers, including Digital Elevation Model (DEM) and Normalized Difference Vegetation Index (NDVI) maps. These datasets were integrated with political and administrative base maps to create an ecological profile of the area, while support from local government units provided additional baseline information on current resource conditions. Mapping and analysis were carried out using Quantum GIS (QGIS) and Google Earth Pro.

2.1 Study area

Surigao del Norte province is in the Northeastern part of Mindanao between 125°15' to 126°15' east longitudes and 9°18' to 10°30' north latitude. It is bounded on the north and east by the Pacific Ocean. The province has abundant mineral reserves including gold, iron, manganese, silica, cobalt, copper, chromite, limestone, silver and among the world's largest nickel deposits. The province falls under the tropical climate type and prone to brief afternoon downpours and thunderstorms.

To clearly define the geographical extent of the study area. This includes determining the boundaries and characteristics of the forested regions to be mapped. Forest cover and freshwater resources of the selected barangays from the three (3) islands of Surigao City were the sampling sites of this study. These islands include Nonoc, Hikdop and Bayagnan. Figure 1 shows the map of the study area.

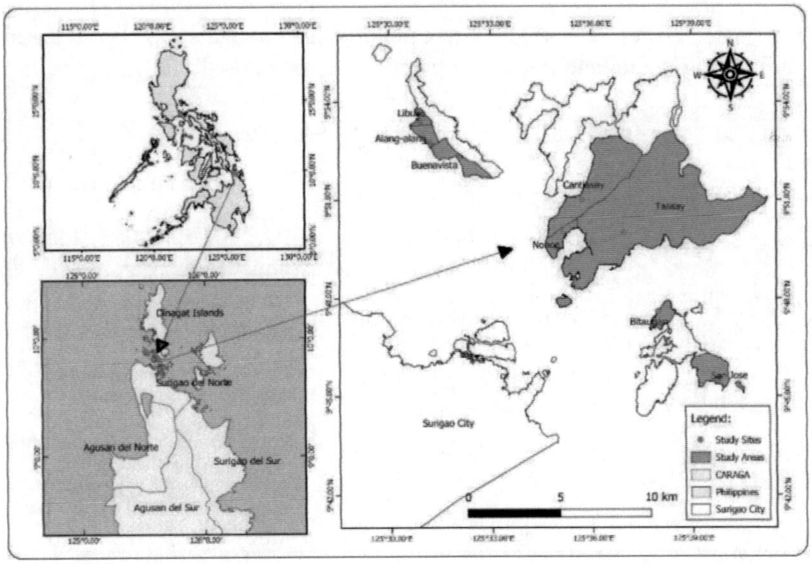

Fig. 1. Location map of the study area.

2.2 Map generation using GIS

2.2.1 Terrain Analysis

To effectively analyze the terrain, several maps were generated using GIS tools, providing valuable insights into how topography influences forest cover and freshwater resources in the island barangays of Surigao City. DEM map was created to display elevation data in 3D, which was useful for examining slope, aspect, and landforms. From the DEM, a slope map was derived to highlight areas with either steep or gentle slopes, an important for erosion and watershed studies. In addition, an aspect map was produced to determine the direction that slopes face, such as north or south, which is particularly relevant for analyzing the climate patterns and vegetation distribution.

2.2.2 Vegetation Analysis using NDVI

To classify and delineate vegetation cover, the NDVI was applied. This index is a widely used spectral indicator of vegetation health and density. It differentiates land cover types by measuring the contrast between red light which vegetation absorbs during photosynthesis and near-infrared light (NIR), which vegetation strongly reflects.

The NDVI formula is expressed as:

$$NDVI = \frac{(NIR - Red)}{(NIR + Red)} \quad (1)$$

For Landsat 8 imagery, the equation becomes:

$$NDVI = \frac{(B5 - B4)}{(B5 + B4)} \quad (2)$$

The computed NDVI values were then categorized into vegetation classes based on their corresponding ranges and land cover interpretation as shown in Table 1.

Table 1. NDVI Classification of Vegetation Cover.

NDVI Range	Color on Map	Vegetation / Land Cover Type	Description
< 0	Light Pink	Water bodies / Non-vegetated	Rivers, seas, lakes, and barren non-vegetated zones
0.0 – 0.2	Pink-Red	Barren / Built-up areas	Settlements, exposed soil, rocky areas, or sand
0.3 – 0.5	Light Green	Moderate vegetation	Grasslands, shrubs, croplands, or secondary growth
0.6 – 0.8	Dark Green	Dense forests / Plantations	Healthy tree cover, plantations, and mature forests
> 0.8	Blue	Very vigorous vegetation	Extremely healthy, dense, and highly productive vegetation

3 Results and discussion

3.1 Profiling of forest cover and freshwater resources

The forest cover profiling was conducted in an area that has identified freshwater source. Among the barangays in each islands, there are three in Hikdop, three in Nonoc, and two in Bayagnan have freshwater source. Presented in Table 2 the data for profiling forest cover and freshwater resources in the island barangays of Surigao City.

Table 2. Forest Cover and Freshwater Resources Profile.

Barangay	Latitude	Longitude	Vegetation	Water sources	Land area (ha)	Elevation (m)	Forest Cover
Hikdop Island							
Alang-alang	9°52.525'	125°31.397'	Tertiary Growth	Stream	107.98	26	Antipolo, Ferns, Cogon grass, Bamboo, Rattan, Banana, Breadfruit, Shrubs
Buenavista	9°52.293'	125°32.095'	Tertiary Growth	Deep Well	217.01	27	Coconut tree, Banana, Grass
Libuac	9°53.400'	125°30.911'	Tertiary Growth	Spring	63.94	21	Grass, Coconut trees, Banana, Cogon grass
Nonoc Island							
Cantiasay	9°51.000'	125°35.714'	Tertiary Growth	Stream	1,030.51	37	Ferns, Screwpine, Fern tree, Cogon grass, Rattan, Coconut trees, Banana
Nonoc_ Jetty3	9°49.985'	125°35.136'	Tertiary Growth	Stream	145.93	17	Fern tree, Ferns, Screwpine, Agoho tree, Shrubs, Cogon grass
Nonoc_ Stream	9°49.571'	125°34.960'	Tertiary Growth	Stream	145.93	24	Pitcher plant, Ferns, Screwpine
Talisay	9°49.998'	125°36.962'	Tertiary Growth	Stream	4,000.06	56	Fern tree, Ferns, Screw pine, Banana, Palm tree
Bayagnan Island							
Bitaugan	9°47.608'	125°37.977'	Tertiary Growth	Deep Well	119.19	27	Grass, Coconut trees, Banana, Cogon grass
San Jose	9°45.487'	125°39.774'	Tertiary Growth	Deep Well	1580.91	67	Trees, Banana, Grass, Shrubs, Ferns

The profiling of forest cover and freshwater resources in the island barangays of Surigao City reveal significant insights into the relationship between land use, vegetation, and water availability. Across the three island groups—Hikdop, Nonoc, and Bayagnan—the dominant vegetation type is tertiary growth, which indicates ecological recovery after past disturbances such as logging, slash-and-burn farming, or land conversion. In Hikdop Island, the vegetation is diverse and includes Antipolo, ferns, cogon grass, bamboo, rattan, banana, breadfruit, and shrubs, reflecting a mix of natural regrowth and agroforestry. Nonoc Island, on the other hand, is notable for its richer biodiversity with species such as fern trees, screwpine, agoho trees, pitcher plants, and palms, which suggests higher conservation value. Bayagnan Island is more grassland and coconut dominated, with scattered bananas and shrubs, highlighting stronger agricultural influence and human intervention.

Freshwater resources vary across the islands and are closely tied to the state of forest cover. In Hikdop and Nonoc, streams and springs are the main sources of water, showing that natural catchment systems remain functional but are vulnerable to deforestation and sedimentation. In contrast, Bayagnan relies heavily on deep wells, pointing to greater dependence on groundwater extraction due to limited surface water availability. Land area and elevation further shape the ecological profile of these barangays. The largest areas, such as Talisay in Nonoc (4,000.06 ha), San Jose in Bayagnan (1,580.91 ha), and Cantiasay in Nonoc (1,030.51 ha), also host denser vegetation and higher elevations, making them critical for watershed and aquifer recharge. Smaller and lower-elevation areas, like Libuac and Nonoc_Jetty 3, are more exposed to risks such as saltwater intrusion into wells and flooding.

This results conformed in the study where streams and rivers that has dense forest cover and less agriculture had better ecological status, those with over 20% disturbances showed poor conditions [9]. The worries about lack of water supply and damage to freshwater ecosystem are increasing, however, studies cannot substantiate the effectiveness of the forest cover in protecting the rivers and streams and reducing the negative impact. Land cover can also be a factor of the water quality of the river and stream [10]. Any disturbances or human activities that could affect the forest cover along the streams and rivers can impair the natural support system of the aquatic organisms and well as can affect the amount and quality of water [11] thus, the community that is dependent in the river or streams as their water supply could also be affected.

3.2 GIS-Based Mapping

3.2.1 Water Sources Map

The water sources map (Figure 2) illustrates the precise locations of streams, wells, and springs in the study area using GPS and valid secondary data, with GIS mapping applied as a cost-effective, reliable, and widely recognized method for water resource management.

Fig. 2. Water Sources Map.

3.2.2 Forest Cover Zoning Map (NDVI based)

Figure 3 presents the NDVI map, derived from satellite imagery, showing the spatial distribution of vegetation health and density across the island barangays of Surigao City. The color-coded legend illustrates five land cover classes: light pink represents water bodies or non-vegetated areas, pink-red denotes barren or built-up surfaces with sparse vegetation, light green indicates moderate vegetation such as grasslands and croplands, dark green corresponds to dense vegetation such as forests and plantations, and blue highlights zones of very vigorous vegetation, typically healthy and highly productive ecosystems.

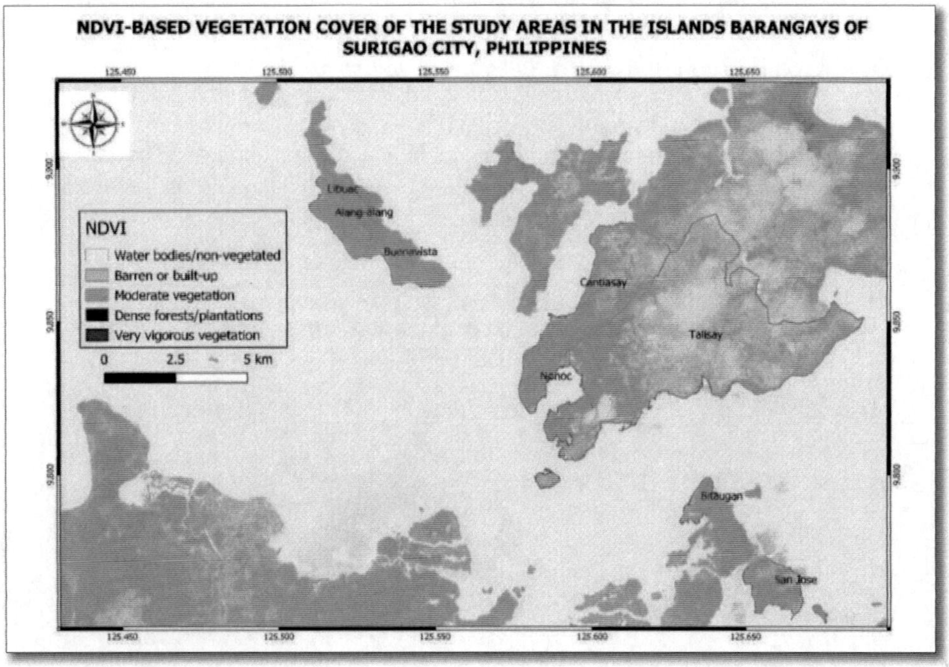

Fig. 3. NDVI Map.

The NDVI distribution reveals distinct vegetation patterns across the study area. Dense and healthy vegetation (dark green) is evident in large portions of Brgys. Nonoc, Cantiasay, and Libuac, reflecting intact forest stands and plantation areas. Very vigorous vegetation (blue) appears in scattered pockets within Nonoc and Cantiasay, signaling well-preserved ecosystems with high biomass productivity. Moderate vegetation (light green) is widespread in Buenavista, Bitaugan, San Jose, and Alang-alang, consistent with grasslands, agricultural fields, and regenerating secondary growth.

In contrast, barren or built-up zones (pink-red) are most noticeable in Talisay, San Jose, and parts of Bitaugan, suggesting land conversion, settlement expansion, or degraded ecosystems. Light pink areas distinctly mark surrounding coastal waters and non-vegetated surfaces, separating marine and terrestrial ecosystems.

The NDVI map indicates that while substantial vegetation cover persists particularly in Nonoc, Cantiasay, and Libuac, there are clear signs of vegetation stress and land-use change in Talisay, San Jose, and Bitaugan. These findings highlight the need for sustainable land management practices, reforestation programs, and conservation measures to protect the ecological integrity of the island barangays and ensure the long-term resilience of their natural resources.

The utilization of NDVI classified forest cover in North China region using Landsat ETM+ data [12]. The analysis revealed that NDVI produced consistent results across the study periods. In contrast, utilization of Google Earth Engine encountered challenges in accurately determining forest cover rates [13]. One major difficulty was the uneven distribution of trees in the study area, particularly in transition zone between forest and grassland. Meanwhile, the use of multiple data sources and approaches in forest cover mapping [14]. Their findings indicated that relying on multi-source data could introduce significant uncertainty, potentially affecting the reliability of the results. Consequently it is suggested that GIS-based studies avoid the use of multi-source datasets to minimize inconsistencies and ensure more accurate outcomes.

3.2.3 Topographic Characteristics and Terrain Features

This study presents terrain maps of the island barangays in Surigao City, Philippines, illustrating the physical features of the landscape such as elevation, slope, and aspect through contouring, shading, and color gradients, which are essential for land use planning, disaster risk management, and ecological assessment, see Figure 4.

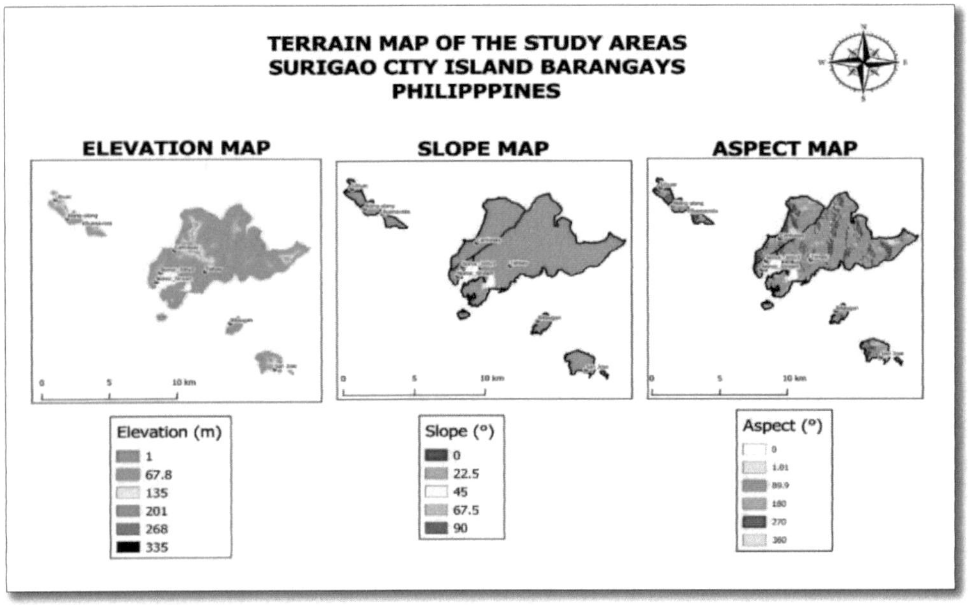

Fig. 4. Terrain Map.

In the elevation map, the elevation ranges from 1 to 335 meters, with the lowest elevations in coastal areas and the highest observed in the central and eastern sections of Nonoc Island and Bayagnan Island. Barangays such as San Jose and Talisay show substantial areas in orange and brown tones, indicating elevations above 200 meters, which can affect accessibility, water drainage, and vegetation types. Alang-alang, Buenavista, and Libuac on Hikdop Island are in the lower elevation category, making them potentially more vulnerable to coastal flooding and sea level rise.

In the slope map, most barangays fall within low slope values (0° to 22.5°), indicating relatively flat or gently sloping terrain that is well-suited for settlement, farming, and basic infrastructure. Steeper slopes are limited or scattered, commonly found in inland or forested areas, such as Talisay and San Jose, where they may pose risks of erosion or landslides during heavy rainfall.

In the aspect map, it shows the directional orientation of slopes, which can influence microclimate, soil moisture, and vegetation. The dominant aspect varies, with a mix of northern (0°), eastern (90°), southern (180°), and western (270°) orientations. The diversity in aspect across Nonoc and Bayagnan suggests varied exposure to sunlight, which can be crucial for agriculture, forestry, and biodiversity management.

The combination of low slopes and moderate elevations in most barangays suggests that these areas are suitable for development, but attention must be paid to watershed protection and erosion control in steeper zones. The aspect variations may help in strategic planting (e.g., shade-loving vs sun-loving crops) and understanding localized weather impacts. These terrain characteristics support site-specific environmental planning, especially for sustainable resource management, hazard mitigation, and climate-resilient development.

The use of DEM has proven to be highly effective for terrain analysis. DEMs are widely applied in assessing various terrain attributes, including slope, slope type, watershed boundaries, and standard flow paths. A comparison accuracy and robustness [15] was performed by placing three different objects on three actual slopes for traversability analysis. This approach effectively demonstrated the capacity of DEMs to illustrate diverse terrain conditions, including cluttered obstacles, potholes, and varying slope gradients, further confirming their reliability and practical value in terrain studies.

4 Conclusion

The utilization of GIS in mapping and analyzing the forest cover and freshwater resources in the island barangays of Surigao City was successfully carried out in this study. Insights into the vegetation distribution, water sources and terrain features were offered in detail through GIS-based approach with the integration of DEM data and field surveys. The landscape in the areas of the study were predominant with tertiary growth forest and the primary source of freshwater are streams, springs, and deep wells. There is a need to focus on conservation efforts as indicated in the NDVI analysis. The Barangays Nonoc, Cantiasay, and Libuac are highlighted to have dense vegetation, while Barangays Talisay, San Jose, and Bitaugan are the areas with vegetation stress and land-use change. Some possible environmental hazards including erosion and flooding were also revealed in the terrain analysis, emphasizing the significance of sustainable land management techniques. Policymakers, environmentalists, and local communities can all benefit from the incorporation of GIS technology, which promotes sustainable development and evidence-based environmental planning. The study underlines the requirement of integrated GIS monitoring systems, sustainable water resource management, reforestation activities, and catastrophe risk reduction techniques to ensure the long-term resilience of these island ecosystems.

References

1. United Nations World Water Assessment Programme, World Water Development Report 2019: Leaving No One Behind, United Nations (2019)
2. Intergovernmental Panel on Climate Change (IPCC), Climate Change 2014: Impacts, Adaptation, and Vulnerability. Contribution of Working Group II to the Fifth Assessment Report of the Intergovernmental Panel on Climate Change, Cambridge University Press (2014)
3. World Health Organization, Progress on Drinking Water, Sanitation, and Hygiene: 2000-2017: Special Focus on Inequalities (2019)
4. Sustainable Development Goals, Department of Economic and Social Affairs, United Nations. https://sdgs.un.org/goals (2020)

 https://doi.org/10.1051/e3sconf/202669001001

5. Harmonized National R&D Agenda 2022-2028, Department of Science and Technology (DOST) https://www.dost.gov.ph/phocadownload/Downloads/Resources/Quick_Links/HNRDA_2022-2028.pdf (2022)
6. M. C. Hansen, P. V. Potapov, R. Moore, M. Hancher, S. A. Turubanova, A. Tyukavina, D. Thau, S. V. Stehman, S. J. Goetz, T. R. Loveland, A. Kommareddy, A. Egorov, L. Chini, C. O. Justice, J. R. G. Townshen, *High-resolution global maps of 21st-century forest cover change*. Science, **342**(6160), 850-853 (2013)
7. B. Lehner, C. R. Liermann, C. Revenga, C. Vörösmarty, B. Fekete, P. Crouzet, P. Döll, M. Endejan, F. Frenken, J. Magome, C. Nilsson, J. C. Robertson, R. Rödel, N. Sindorf, D. Wisser, *High-resolution mapping of the world's reservoirs and dams for sustainable river-flow management*. Frontiers in Ecol. and the Env. **9**(9), 494-502 (2011)
8. M. Bonell, L. A. Bruijnzeel (Eds.). *Forests, Water and People in the Humid Tropics: Past, Present and Future Hydrological Research for Integrated Land and Water Management.* Cambridge University Press (2005)
9. M. Tolkkinen, S. Vaarala, J. Aroviita, *The importance of riparian forest cover to the ecological status of agricultural streams in a nationwide assessment*. Water Res. Mngt, **35**(12), 4009-4020 (2021).
10. D. Brogna, A. Michez, S. Jacobs, M. Dufrêne, C. Vincke, N. Dendoncker, *Linking forest cover to water quality: A multivariate analysis of large monitoring datasets.* Water, **9**(3), 176 (2017)
11. L. E. England, A. D. Rosemond, *Small reductions in forest cover weaken terrestrial-aquatic linkages in headwater streams.* Freshwater boil., **49**(6), 721-734 (2004)
12. K. Jia, S. Liang, L. Zhang, X. Wei, Y. Yao, X. Xie, *Forest cover classification using Landsat ETM+ data and time series MODIS NDVI data*. Intl. J. Appl. Earth Observ. and Geoinfo, **33**, 32-38 (2014)
13. X. Zhang, T. Long, G. He, Y. Guo, R. Yin, Z. Zhang, B. Cheng, *Rapid generation of global forest cover map using Landsat based on the forest ecological zones*. J. Appl. Remote Sensing, **14**(2), 022211-022211 (2020)
14. Y. Qin, X. Xiao, J. Dong, G. Zhang, M. Shimada, J. Liu, B. Moore III, *Forest cover maps of China in 2010 from multiple approaches and data sources:* PALSAR, Landsat, MODIS, FRA, and NFI. ISPRS J. Photogrammetry and Remote Sensing, **109**, 1-16 (2015)
15. X. Meng, Z. Cao, S. Liang, L. Pang, S. Wang, C. Zhou, *A terrain description method for traversability analysis based on elevation grid map*. Intl. J. Adv. Robotic Sys, **15**(1), 1729881417751530 (2018)

Afforestation of tropical watersheds to minimize flood risk in the parish of Valle Hermoso, Ecuador

Patricio Jiménez-Pozo[1], *Eduardo* Argotti-Valencia [1], *Darwin* Suasnavas[1] and *Hugo* Bautista[1,2*]

[1]Armed Forces University, Quito, Ecuador
[2]Kazan Federal University, Kazan, Russia

Abstract. This paper presents the implementation and outcomes of a community-based afforestation project in the Valle Hermoso parish (Santo Domingo de los Tsáchilas, Ecuador), aimed at mitigating flood risk. From 2021 to 2024, the University of the Armed Forces (ESPE) facilitated the distribution of over 2,500 Guadua angustifolia Kunth (caña guadua) plants to 250 farms across nine sub-districts, engaging 230 volunteers from the university and local community. The study maps historical flood zones and the newly forested areas surrounding the Blanco and Cristal rivers. After 3.5 years, a 55% survival rate indicates successful plant establishment and a positive project outcome. The afforestation has demonstrably increased dry-season stream flow, reduced flooding, and stabilized riverbanks, thereby minimizing landslides. These results position the project as an effective and scalable model for the broader Tsáchila province, though scalability may encounter financial, logistical, and managerial hurdles. The primary challenge remains ensuring long-term sustainability, which is heavily dependent on continued community participation and support, factors vulnerable to external economic and social pressures.

1 Introduction

Flooding is a recurring and devastating natural hazard affecting communities worldwide. Tropical regions, characterized by intense rainfall and steep terrain, are particularly vulnerable. During the last 40 years, the Pacific coast of South America is more prone to severe drought, particularly in Chile, while Peru and Ecuador experience floods, mudslides and landslides during El Niño events [1,2]. Ecuador's rainy season varies by region, but the coastal and Amazon regions typically experience rain from December–May. March and April are the wettest months, so precipitations are more likely than usual in all regions of the country [1]. The most affected provinces are El Oro, Esmeraldas, Guayas, Los Ríos, Manabí, Santa Elena and Santo Domingo de los Tsáchilas. This last province is located in the equatorial zone of Ecuador, which is characterized by high humidity and high rainfall. Intensive deforestation has led to the disruption of the region's ecosystem, a decrease in the

* Corresponding author: hbautista@mail.ru

© The Authors, published by EDP Sciences. This is an open access article distributed under the terms of the Creative Commons Attribution License 4.0 (https://creativecommons.org/licenses/by/4.0/).

soil's ability to retain moisture, and an increased risk of catastrophic natural events such as floods and droughts. One striking example of intensive deforestation in Santo Domingo de los Tsáchilas is the construction project of the El Corazón road, which led to significant destruction of the tropical forest. Another example is the development of a gold mine within the Cotacachi-Cayapas National Park, which caused serious environmental problems [3,4]. Conservación Internacional Ecuador y Forest Trends. Quito, Ecuador). Valle Hermoso parish, is an important agricultural zone inside the mentioned province, who needs to combating these threats and maintaining environmental sustainability.

Guadua bamboo species, and specifically caña guadua (Guadua angustifolia), is known for its ability to retain water and improve soil health. The principal characteristic it his dense root system, working as a sponge, absorbing and storing water, which helps reduce runoff and replenish aquifers. This water retention capacity is helps soil stabilization and erosion prevention.

Purpose and Objectives of the Study

The ESPE Armed Forces University- branch Santo Domingo, launched a 36-month outreach project starting in September 2021. The plan focuses on watershed afforestation to mitigate flooding. To this end, partnerships were established between ESPE and the parish population. Nine of the 21 subdistricts (Tsachila province) participated, working on 250 farms, benefiting more than 800 people directly and approximately 4,300 people indirectly.

The main objectives were:

- Analysis of flood scenarios over the last 40 years
- Afforestation activities with the locals.

This research proposes afforestation, the intentional planting of caña guadua on previously deforested or degraded land, as a viable and sustainable strategy to mitigate flood risk in Valle Hermoso. Guadua plants, particularly native species, play a crucial role in regulating the hydrological cycle, enhancing soil stability, and reducing the impact of heavy rainfall. By restoring natural cover withing the watershed, we aim to improve its natural capacity to absorb and store water, thereby reducing the magnitude and frequency of flood disasters.

2 Metodology

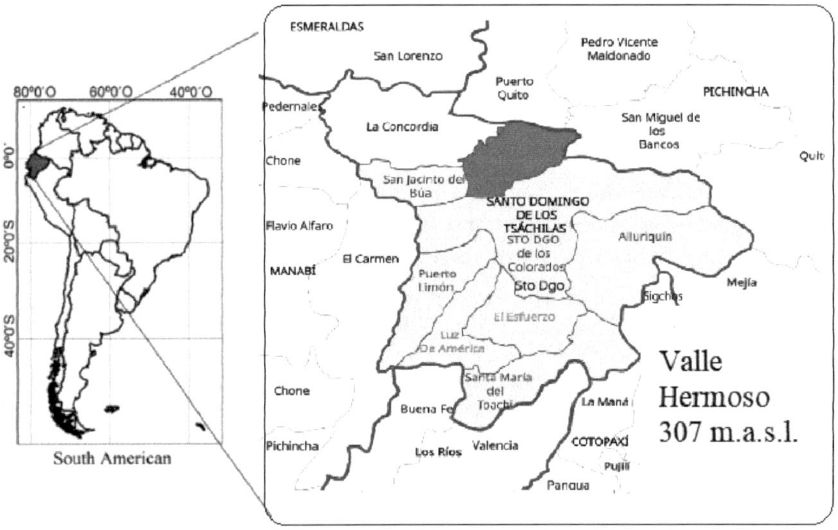

Fig. 1. Valle Hermoso parish, Tsachila province, Ecuador.

Valle Hermoso is a rural parish located north of the Santo Domingo de los Tsáchilas province (Fig.1), at Km. 25 on the Santo Domingo-Quinindé highway. It is bordered by two rivers: the Blanco River and the Cristal River. Rainy weather causes the riverbeds of these two rivers to collapse and rise above their normal levels, flooding the town and causing adverse effects, primarily to more than fifty families living in the area [5,6].

To comprehensively address the set goals of mitigating flood risk through afforestation in Valle Hermoso, this research implemented a multifaceted approach. Firstly, we employed dynamic modeling techniques to simulate the water flood behavior across the territory. This involved analyzing historical rainfall patterns, topographic data, and existing land cover to create a predictive model of flood inundation areas (Fig.2).

Secondly, the research focused on afforestation using guadua bamboo, a species known for its rapid growth, strong root system, and ability to stabilize soils. Afforestation efforts were coupled with rigorous field monitoring and data collection across the afforested areas and the surrounding watershed. The planting density of guadua (Guadua angustifolia) varies depending on the purpose of the plantation. For erosion control, a high density (e.g., 3x3 meters) was used. In general, a planting density of between 278 and 625 plants per hectare can be considered, depending on soil characteristics[7,8]. This monitoring encompassed several key parameters, including the vegetation state of the guadua bamboo stands (e.g., growth rate, density, health), the condition and availability of water resources (e.g., streamflow, groundwater levels, water quality), and the overall state of the land (e.g., soil erosion, soil moisture, sediment deposition). The gathered data served to track the progress of the afforestation project and to quantify its effect on hydrological processes over time.

Finally, the research sought to identify the critical success factors essential for implementing sustainable and effective afforestation projects in tropical environments. This component involved reviewing existing literature on afforestation best practices, conducting interviews with local stakeholders (e.g., farmers, community leaders, government officials), and analyzing the lessons learned from the guadua bamboo afforestation initiative. By pinpointing these key factors, we aim to provide practical guidance for future afforestation endeavors in Valle Hermoso and similar tropical regions, maximizing the likelihood of achieving long-term flood mitigation and ecosystem restoration benefits.

3 Results and discussion

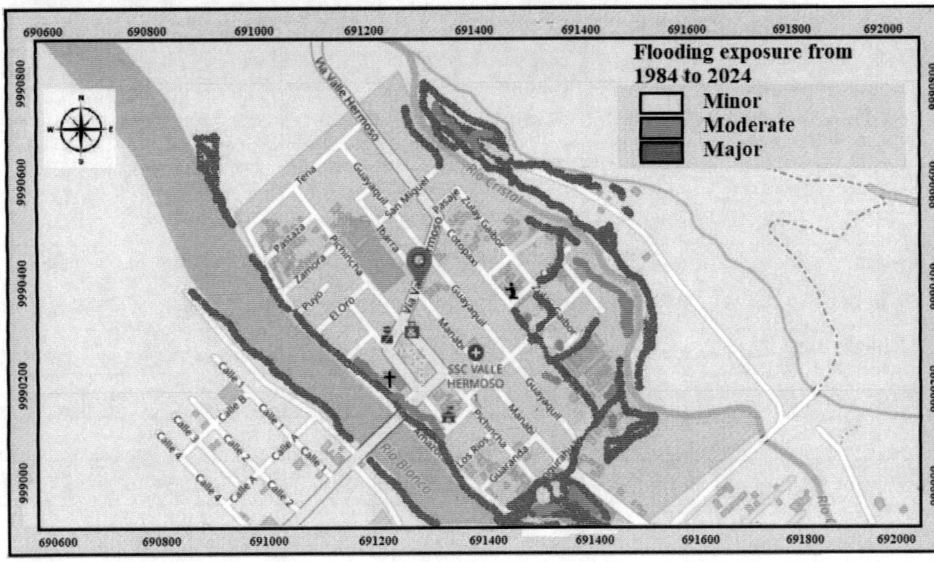

Fig. 2. Valle Hermoso's flooding impacts during severe rainfalls from 1984

The main causes of forest degradation are agricultural activities, timber harvesting and urbanization. These processes have led to the loss of about 70% of the original forest massif, an increase in the area of ravines and a deterioration in the condition of surface waters [9,10,11].

After three years, we can say that afforestation with cana guadua plants significantly reduce the risk of extreme weather conditions and increase the stability of watersheds (Tab.1). Cana guadua can retain rainwater, reducing runoff and promoting moisture accumulation in the soil (Fig.3). This helps improve the local microclimate and ensures a sustainable water supply to populated areas.

Fig. 3. Flood reduction function of caña guadua on their interception evaporation and root's storage capacities.

Table 1. Growth of caña guadua

Clump age (months)	Total culms	Thickest culm (cm)	Tallest culm (m)
Planted 9/01/2021	8	0.7	1.05
6	12	1.7	1.7
12	14	4.3	4
18	16	7.3	7.1
+24	24	9.0	9.0

Economic benefits of the forest project

Implementation of the forest restoration project brings a number of economic benefits:

- Increased agricultural productivity due to stable water supply.
- Development of ecological tourism associated with the uniqueness of the local flora and fauna.
- Improvement of public health due to a reduction in the number of diseases associated with a lack of clean drinking water.

4 Conclusions and recommendations

The flow was reduced by afforestation during the wet season, indicating a reduction in flood risk. The water input to land during the dry season, i.e., the sum of dry-season precipitation and the carryover of land water, increased due to warming, albeit only slightly. However, these increases did not lead to an increase in flow (i.e., an increase in water resources) during the dry season because evapotranspiration increased significantly during this period. The decrease in high flow in September due to afforestation indicated a reduction in flood risk, whereas the decrease in low flow during the dry season due to the expansion of afforested areas indicated an exacerbation of water scarcity risk.

Climate warming exacerbates the risks of annual water resource shortages, flooding during the wet season, and water scarcity during the dry season. Afforestation can exacerbate the risks of annual water resource shortages and water scarcity during the dry season, while it may reduce the risk of flooding during the wet season. Since reducing flood flows would not reduce annually available water resources, this effect would be considered beneficial. It is important to note, however, that even achieving 50% afforestation is extremely challenging in practice since many people rely on income generated through work on cultivated lands. Therefore, a 100% afforestation scenario is practically impossible to achieve. In conclusion, the hydrological effects of afforestation that reduce flood risks but exacerbate drought risks must be equally considered in regions exposed to both hazards, such as the White River or Crystal River basins.

The authors thank the residents of Valle Hermoso for their enthusiasm and availability to develop the project. We also thank SENECYT of Ecuador for their valuable contribution to the scholarship recipients.
This paper has been supported by the Kazan Federal University strategic academic leadership program (PRIORITY-2030).

References

1.Ecuador Systematic Country Diagnostic. *In World Bank*, Washington, DC, 2018.eBooks. doi: 10.1596/30052.

2. Glantz, M.H. *Currents of Change: El Niño's Impact on Climate and Society*. Cambridge University Press, Cambridge, 2001 **84**:p.1–252.

3. Sierra, R. *Patrones y factores de deforestación en el Ecuador continental, 1990-2010. Y un acercamiento a los próximos 10 años.* Conservación Internacional Ecuador y Forest Trends, Quito,Ecuador, 2013. **15**:p.57.

4. Bautista, H., Moreno, M., and Argotti, E. *Reforestation and its role in sustainable development of one community of Luz de América, Ecuador, with eight subtropical plant species implemented*. E3S Web of Conferences, 2024. **42**:p.537.

5. Morillo, J. A. G., Guamán, M. A. R., & Franco, P. S. *Modelamiento hidráulico del río Blanco y el río Cristal, para el análisis de inundaciones en la cabecera parroquial de Valle Hermoso, cantón Santo Domingo*. Sinergia Académica, 2023. **6**:p.166–183.

6. Decentralized Autonomous Government of Santo Domingo-Ecuador GAD, 2016: Forest restoration of the province. *Data on forests and ecosystems of continental Ecuador*, 2016.

7.Somasekhar, P.V., Lakshmikanth, R.N., Rathore, T.S., Reddy, K.S. and Lakshman, A.C. *Macro-propagation of Guadua angustifolia Kunth. – An exotic and fast growing bamboo species*, Indian Forester, 2004. **130**:p.655-662.

8. Moosa, M., P, A. C., M, A., B, S. V., & Viswanath, S. *A review of the current state of knowledge on Pseudoxytenanthera ritchiei- a lesser known bamboo species with high utilization potential from Peninsular India*. Journal of Bamboo and Rattan, 2023. **21**(4), p.146–165.

9. Kosugi K.I. *Effects of soil layer and permeable bedrock on flood and base flow discharges from a forested watershed.* Journal of Japan Society of Hydrology and Water Resources, 2007. **20**:201, p.13.

10. Kleemann, J., Zamora, C., Villacis-Chiluisa, A.B., Cuenca, P., Koo, H., Noh, J.K., Fürst, C., Thiel, M. *Deforestation in Continental Ecuador with a Focus on Protected Areas*. Land, 2022. **11**(2), p. 268

11. Shabalina, S. A., Bautista, H., Ulengov, R. A., Mustafin, M. R. *Theory Of Compromise Solutions As Effective Tool For Improving Regional Tourism Sphere*. European Proceedings of Social and Behavioural Sciences, 2022. **125**:p. 917-923.

Review of Strategies to Control Air Pollution due to PFAS

Prashant Rajurkar[1*]

[1]Department of Environmental Affairs, HF Sinclair, Dallas, TX, USA

Abstract. Per and polyfluoroalkyl substances (PFAS), also known as "forever chemicals" have emerged as the next environmental threat due to their persistence and challenges in capture and control. The water contamination aspect of the PFAS has been extensively researched, however, the emissions into the atmosphere are a growing concern due to their volatility, long-range transport potential, and resistance to degradation. As a result of discovery of air pollution pathways of PFAS due to industrial processes, consumer usage and waste treatment, researchers have been identifying innovative technologies to mitigate the PFAS related air pollution. This study reviews various types of existing technologies such as oxidation, absorption and adsorption. In addition, a review of emerging technologies such as non-thermal plasma treatment was also investigated and presented in this study. Key research gaps and opportunities for future innovation were identified and possibilities of multi-prong approach were investigated to mitigate PFAS associated air pollution.

1 Introduction

PFAS are a large group of fluorinated organic chemicals, frequently referred as "forever chemicals" are used extensively due to their thermal and chemical stability. The history of PFAS goes back to 1940s when these chemicals were developed and used in Teflon®, firefighting foams and stain-resistant products. Around 2000s, the US Environmental Protection Agency (EPA) began investigating PFAS contamination triggered by internal documents from 3M and DuPont and water contamination findings in West Virginia. By 2002, 3M voluntarily phased out perfluorooctane sulfonic acid (PFOS) in the US and in 2009, the PFOS and perfluorooctanoic acid (PFOA) were added to the Stockholm Convention on Persistent Organic Pollutants [1] for global regulation and the US EPA issued the provisional health advisories. In 2016, the US issued PFOA and PFOS Drinking Water Health Advisories [2]. In 2021, The EPA Council on PFAS was established with a mandate to develop a wholistic approach to protect the public health and the environment from the impacts of PFAS. The PFAS Council developed the PFAS Strategic Roadmap [3]

While there was a significant amount of research on PFAS impacts associated with water, the air impact pathways and effects were not known until 2001 when PFAS was suspected in the ambient air near the DuPont facility in Parkersburg, WV. Between 2005-2007, studies

* Corresponding author: prashant.rajurkar@hfsinclair.com

© The Authors, published by EDP Sciences. This is an open access article distributed under the terms of the Creative Commons Attribution License 4.0 (https://creativecommons.org/licenses/by/4.0/).

began identifying fluorotelomer alcohols (FTOHs) and sulfonamides in indoor and outdoor air. These volatile category precursors are suspected to transform into persistent PFAS in the environment and can travel long distances. During the years 2008 – 2010, PFAS was detected in Arctic and remote locations thereby proving the long-range atmospheric transportation characteristics [4]. By 2015, US EPA formally established the air-to-soil and air-to-water pathways. The awareness about airborne PFAS and associated impacts has evolved since then and lead to the establishment of the Office of Air and Radiation by the US EPA PFAS Council.

2 Sources of PFAS Air Emissions

In order to understand the control and mitigation strategies, it would be prudent to review the sources of the PFAS species. Like other air pollutants, the airborne PFAS can originate from a variety of point and area sources. The following are some of the known PFAS sources:

- Industrial Manufacturing and Processing
  - Fluoropolymer synthesis: Polymerization can release volatile PFAS (PFAAs, FTOHs, sulfonamide precursors).
  - Surfactants/repellents production: Dryers, exhaust stacks, and formulation tanks in textiles, carpets, and paper manufacturing emit mixed PFAS species.
  - Fluorochemical handling: Blending, milling, and transfers generate fugitive emissions from powders and liquids.
- Thermal Treatment of Waste
  - Hazardous waste incineration: Insufficient temperature or residence time leads to partial PFAS destruction, forming ultrashort-chain acids and byproducts (e.g., HF, COF_2).
  - Sludge drying & biosolids incineration: PFAS in municipal/industrial sludge volatilize during drying/combustion.
  - Thermal desorption units (TDUs): Incomplete destruction can volatilize PFAS precursors from contaminated soils.
  - Metal smelting & aluminum casting: Fluorinated fluxes/coatings release PFAS vapors during melting and burnout.
- Product Use and Degradation
  - Use-phase volatilization & weathering: Heat, abrasion, and UV promote release of precursors (e.g., FTOHs) from treated products; atmospheric oxidation converts these to terminal PFAAs.

3 Fate and Transport of PFAS in the Atmosphere

The physical forms in atmosphere, transformation, transport mechanism and environmental distribution after deposition constitutes the fate and transport of PFAS in the atmosphere. In the atmosphere, PFAS can occur in gas and particle phases, or other aerosols suspended in the air. The gaseous phase includes volatile and semi-volatile compounds such as FTOHs and perfluoroalkane sulfonamides (FASAs). Ionic PFAS such as PFOA, and PFOS with low vapor pressure and high solubility tend to be the dominant species in airborne particulate matter.

The particle-associated phase can absorb to atmospheric particulate matter such as PM_{10} and $PM_{2.5}$, which results in long-range transport and deposition. The atmospheric half live of precursors such as FTOH of 20 days further facilitates in intercontinental transport [5]. There are documented evidence of PFAS transport from manufacturing regions in China and North America to remote regions such as the Arctic and Tibet, confirming long range atmospheric transport (LRAT).

The particle bound PFAS can settle onto land and water surfaces. In the areas downwind of emission sources, the PFAS can be removed from the air due to rain or snow thereby traveling with precipitation on land or water bodies. It can also bioaccumulate in plants and animals.

The atmospheric transformation includes oxidation of semi-volatile compounds such as FTOHs and FASAs, eventually transforming into terminal PFAS compounds like PFOA and PFOS. Table 1 summarizes the fate and transport of PFAS in the atmosphere.

Table 1. Summary of Fate and Transport of PFAS in Atmosphere.

Process	Description
Long-Range Transport	PFAS precursors can travel thousands of kilometers via global wind currents.
Photooxidation	FTOHs and FASAs degrade to stable acids such as PFOA or PFNA
Dry Deposition	Particle-bound PFAS settle onto land and water bodies.
Wet Deposition	Rain and snow scavenge PFAS from air, contributing to surface water loading.

The dynamics of atmospheric residence times varying from days to months, enabling regional and global dispersion necessitate effective airborne control strategies at the source and during waste processing.

4 Control Strategies

The control strategies for airborne PFAS emissions includes source control and treatment, identifying alternatives through green chemistry, and establishing a monitoring and risk assessment framework.

4.1 PFAS Air Pollution Control Technologies

4.1.1 Adsorption-Based Techniques

Adsorption based techniques are effective in control PFAS air pollution as volatile and semi-volatile PFAS compounds could be captured from industrial stacks, off-gases from remediation processes and ambient air stream in some cases. This option is comparatively less energy intensive as compared to thermal destruction and a potential for modular deployment in both fixed and mobile applications. The process of adsorption involves passing PFAS molecules over high surface area carbon structures. Granular Activated Carbon (GAC) or Powdered Activated Carbon (PAC) used in the adsorption process is often derived from coconut shell, wood or coal. Given the modular nature of these systems, this technique is usually applied to industrial stacks, air filtration systems and emergency vapor capture systems in PFAS remediation. The key performance factor in adsorption is the PFAS chain length, the longer chain adsorbs more readily as compared to shorter chain molecules. Humidity and co-contaminants can also impact the adsorption rates. In addition, the adsorbing media requires proper treatment such as thermal or solvent regeneration prior to end of disposal.

4.1.2 High-Temperature Thermal Oxidation (HTTO)

This technique is used to eliminate PFAS from gas streams. The destruction method involves combustion of PFAS gas stream in a refractory-lined combustion chambers at temperatures above 1100°C [6]. The high combustion temperature is required to break the C-F bonds. Table 2 describes the operating parameters of HTTO process. The following is a general equation that describes the reaction chemistry of HTTO:

$$C_7F_{15}COOH + O_2 \rightarrow 7\ CO_2 + 15\ HF + H_2O \qquad (1)$$

Table 2. Operating Parameters of HTTO.

Parameter	Range	Description
Combustion temperature	>1,100°C (2,012°F)	Required to achieve >99% destruction of stable PFAS
Residence time	~2 seconds	Allows for complete oxidation of fluorinated compounds
Oxygen concentration	6–10% excess O_2	Supports complete combustion; prevents dioxin/CO formation
Quenching system	Rapid post-combustion cooling	Prevents recombination of intermediates
Scrubbing system	Caustic wet scrubber	Captures HF, COF_2, and other acidic byproducts

The destruction efficiency of HTTO has been demonstrated to be greater than 99.99%. Given optimum residence time control and scrubbers post combustion chamber, PFAS could be destroyed with minimal toxic byproduct formation. However, there are some design and operational challenges that are faced by the HTTO space. Some of the waste gases such as HF are highly acidic and requires appropriate material of construction. There could be generation of toxic byproducts if the residence time is inadequate, combustion temperatures are lower and if there is a lack of scrubbing operation post combustion. Waste heat recovery systems could be implemented to leverage the high energy intensity nature of this process. Other strategies to overcome some of the design and operational challenges include:

1. Incorporating pre-treatment such as filtration or gas cooling to address high particulate loading
2. Reaction kinetics could be better controlled if direct or indirect cooling or quenching mechanisms are installed.
3. Ensure scrubbing process and equipment are in working condition to tackle gases such as HF, SO2 and COF2.

4.1.3 Non-Thermal Plasma Reactors

This technology involves high-voltage electrical discharge to generate reactive species such as electrons, ions and radicals. The electrically charged ions and electrons initiate the chemical reaction that can lead to degradation of the PFAS into smaller compounds such as CO2, HF and other short-chain fluorinated species. The destruction efficiency of this technology can range between 60-95% depending upon the target pollutant and reactor design. The energy yield is around 10-100 g PFAS/kWh and the operating temperature is usually below 200 °C. The lower operating temperature is one of the advantages of this technology. Other advantages include avoidance of thermal byproducts such as dioxins and furans. This technology also uses comparatively lower energy as compared to HTTO. The typical residence time ranges between milliseconds to seconds. However, there are certain challenges to this technology including incomplete destruction, scale-up complexity and equipment fouling.

4.1.4 Advanced Oxidation Technologies (AOTs)

Similar to non-thermal plasma technology, the AOT utilizes reactive species, mainly hydroxyl radicals (•OH) and superoxide (O_2^-) to initiate non-selective oxidation of volatile and semi-volatile PFAS. Semiconductor materials such as titanium dioxide, zinc oxide laced with WO3 is activated by UV light to generate electron-hole pairs that create oxidative radicals. A sample reaction is described as follows:

$$ZnO + hv \rightarrow e^- + h+ \quad (2)$$
$$h+ + H2O \rightarrow \bullet OH \quad (3)$$
$$e^- + O2 \rightarrow O_2^- \bullet \quad (4)$$

These species react with PFAS molecules via decarboxylation or defluorination and oxidation of precursors to terminal acids. The destruction efficiency can range between 60-95% for PFOS, PFOA and FTOHs. However, the efficiency is limited for stable acids like PFOS and PFOA due to strong C-F bongs. The efficiencies are improved by adding oxidants along with the UV. This technology has the advantage of operating in mild conditions with minimal thermal footprint. The operating temperature can range from ambient to 120 C. The residence time can range from 0.5 – 3 seconds.

Some of the challenges in this technology include incomplete degradation of stable PFAS. Terminal acids such as PFOA requires enhanced UV power. Photocatalyst fouling can occur with humid or particulate loading which requires pre-filtration. For high-flow systems, light penetration limitation could affect photocatalysis. The short chain PFAS molecules are more mobile, hence they are less reactive under the photocatalysis conditions. However, with doping and integrating with non-thermal plasma or catalytic oxidation, these challenges could be addressed. Table 3 summarizes the above discussed air pollution control technologies for PFAS.

Table 3. Summary of Air Pollution Control Technologies for PFAS.

Technology	Principle	PFAS Removal Efficiency	Consideration
Activated Carbon Adsorption	Physical adsorption	Moderate to High	Depends on PFAS volatility
Ion Exchange Resins	Electrostatic interaction	Moderate	Limited field data
High-Temp Thermal Oxidation	Combustion >1100°C	High (>99%)	Requires scrubber to remove byproducts
Non-Thermal Plasma Reactors	Reactive species formation	Moderate (Pilot-stage)	Under active research
UV-Photocatalysis	Oxidative degradation	Low to Moderate	Effective on PFAS precursors

4.2 Emerging Solutions

As evident from the previous section, the airborne emissions of PFAS are complex to treat and require novel approach. With the intensification of regulatory scrutiny along with new discoveries about the impact from airborne PFAS, emerging technologies are being developed to address this complex challenge. Some of these innovations are further development and efficiency enhancement of non-thermal plasma reactors, catalytic filtration, advanced adsorption and data-driven process optimization. These technologies offer promising pathways to reduce PFAS air emissions and mitigate their environmental and health impacts.

4.2.1 Catalytic Filtration Systems

This emerging air pollution control technology is designed to degrade volatile and semi-volatile PFAS compounds in gaseous streams. The treatment system includes a ceramic or metallic substrate coated with noble or rare earth metal catalysts such as cerium, platinum, or manganese oxides. This system requires operation at moderate to high temperatures (typically 300–700°C). The PFAS-contaminated air passes through the catalytic bed thereby initiating the oxidation reactions that break the strong C–F bonds. As a typical goal of this technology is to convert harmful fluorinated organics into non-toxic byproducts such as CO_2, HF, and H_2O. Unlike traditional thermal incineration, catalytic systems are less energy intensive, produce fewer toxic byproducts, and demonstrate greater operational stability, specifically when integrated with scrubbers to capture acidic gases. Pilot studies have demonstrated destruction efficiencies exceeding 90% for PFAS making catalytic filtration a promising tool for industrial facilities seeking to reduce airborne PFAS emissions.

However, the catalyst deactivation can place due to fouling or fluorine accumulation. The technology requires optimization of residence time and temperature control

4.2.2 Biochar and Functionalized Sorbents

This passive capture technology is generating interest due to lower cost intensity in the removal of volatile and semi-volatile PFAS species from the atmosphere. The biochar is derived from the pyrolysis of organic biomass, and it offers high surface area for the adsorption of PFAS precursors such as FTOHs and FASAs. Research has shown that biochar can be functionalized with amine groups, metal oxides or activated carbon coatings to improve their selectivity and affinity for a larger range of PFAS species [7]. This technology is currently under developmental stage for large-scale operations. The low energy requirements, renewable nature and onsite regeneration potential makes this a lucrative option.

One of the drawbacks of this technology is that it doesn't result in complete destruction, rather the media must be either disposed or regenerated. In addition, the adsorption capacity and efficiency depends on the temperature and chain length of the PFAS

4.2.3 Supercritical Water Oxidation (SCWO)

This is an emerging high-efficiency destruction technology designed to convert PFAS compounds into benign end-products under extreme reaction conditions. The process involves passing contaminated gases through an aqueous phase followed by exposure to supercritical water, with temperatures above 374°C and a pressure of 22.1 MPa, in the presence of an oxidant such as oxygen or hydrogen peroxide [8]. These conditions render the organic compounds fully miscible in water, which facilitates in rapid and complete oxidation reactions. The contaminated vapor streams are usually pre-scrubbed or absorbed into water to facilitate feed into the SCWO reactor. Similar to the other destruction technologies, this process ensures the conversion of persistent fluorinated compounds into CO_2, HF, and H_2O. The bench scale setups have demonstrated destruction efficiencies exceeding 99.99%. SCWO is particularly attractive for treating concentrated off-gas, such as from AFFF destruction or fluorochemical manufacturing, due to its minimal byproduct formation and ability to handle mixed waste streams.

While system complexity, high capital and operational cost, inlet gas pre-treatment requirement and material corrosion remain engineering challenges, SCWO represents a near-zero emissions solution for the complete destruction of PFAS captured from air.

4.2.4 AI-Enhanced Predictive Emission Modeling

The use of artificial intelligence (AI) and machine learning (ML) are increasingly being integrated into predictive emission modeling frameworks to improve the detection, quantification, and management of airborne PFAS. While these tools are not treatment technology, they leverage large datasets—ranging from process conditions, meteorological parameters, and historical emission profiles—to forecast PFAS dispersion patterns, transformation pathways, and potential human or ecological exposure zones. AI models can dynamically simulate the behavior of volatile precursors like FTOHs and FASAs, accounting for their chemical degradation, partitioning, and long-range atmospheric transport. AI models are also being trained on real-time sensor data from facility stacks, ambient air monitors, and meteorological feeds to generate proactive alerts when emission trends deviate from expected baselines. In industrial settings, AI-enhanced systems can guide adaptive process control, optimize scrubbing or capture efficiency, and support regulatory reporting with more precise and validated predictions. By enabling data-driven risk forecasting and emission reduction planning, AI-powered modeling represents a transformative tool in managing the complex and diffuse nature of airborne PFAS pollution [9]. However, this is dependent on high-quality input data and model validation. Table 4 summarizes the emerging PFAS treatment technologies.

Table 4. Summary of Emerging PFAS Treatment Techniques.

Technology	Primary Mechanism	Destruction/Removal Efficiency	Status
Catalytic Filtration	Surface-catalyzed PFAS decomposition	Moderate to High (~80–90%)	Pilot-stage
Biochar & Functional Sorbents	Adsorption via surface functional groups	Moderate (depends on PFAS type)	Field-tested
Supercritical Water Oxidation	Oxidation under high temp/pressure	Very High (>99.99%)	Full-scale trials
AI-Enhanced Emission Modeling	Predictive analytics & dispersion tracking	Indirect – prevention focused	Operational pilots

5 Monitoring and Risk Management Framework

The pathway to develop an effective PFAS management framework includes prevention of uncontrolled PFAS releases to the atmosphere, detection and quantification of PFAS at source and in ambient atmosphere, respond with control technologies, evaluate long-term exposure and policy effectiveness and finally communicate the risks and outcomes to stakeholders. Figure 1 describes the PFAS air pollution control framework.

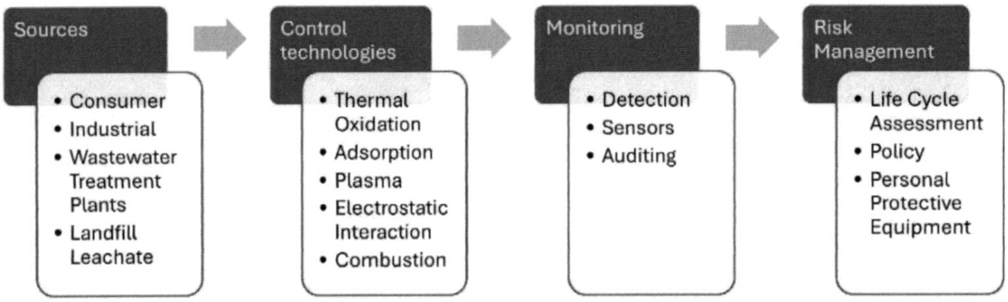

Fig. 1. PFAS Air Pollution Control Framework.

Within the monitoring phase, the detection and quantification of PFAS has been in an ever-evolving phase. The source monitoring framework involves identifying PFAS emissions using stack testing, leak detection, process reviews and inspections. The quantification can be done either using existing methods such as EPA Method OTM-50, TO-134 or TOF-MS, or by engineering estimates. Integrating PFAS sampling with existing sampling ports and real-time gas sensors for PFAS precursors such as FTOHs and PFOS would enhance the quantification. Apart from source monitoring, ambient monitoring using fixed air samplers, passive samplers, mobile monitoring units and real-time analytics assists in further quantification.

The risk management aspect includes managing existing PFAS risks through policy, personal protective equipment and life cycle assessment (LCA). Air filtration systems and personal protective equipment reduce exposure risks in workplaces handling PFAS. Zone monitoring in high-risk areas such as coating lines and thermal treatment zones will further enhance the monitoring network. The LCA evaluations of PFAS assess the total environmental burden across air, water, soil and human health. It also presents the trade-offs in production vs substitution vs treatment. A multi-media LCA model is much more beneficial than a specific media which can result in better informed decision making.

Given the fact that airborne PFAS control is still at its infancy stages, there are several barriers that continue to limit the widespread application and reliability of PFAS air control technologies. These challenges are summarized in the Figure 2.

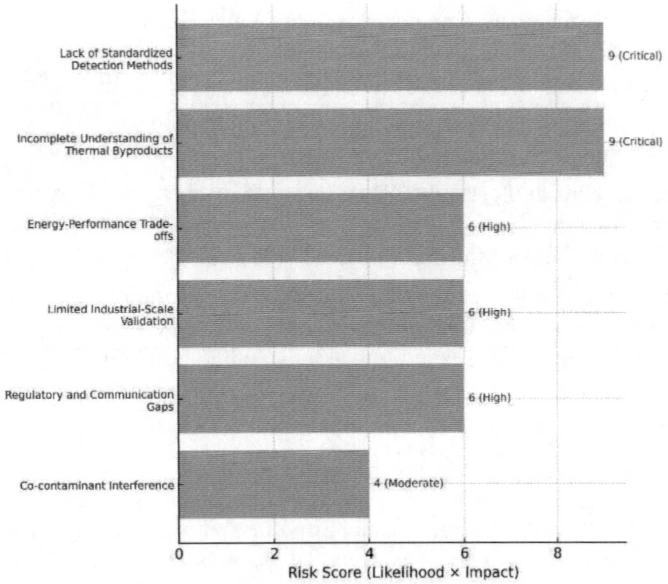

Fig. 2. Risk Ranking of PFAS Air Emission Control Challenges.

6 Conclusion

The PFAS related air pollution control is fairly complex and requires a cross-disciplinary and coordinated approach. The range of PFAS species, their chemical and physical behavior in different atmospheric conditions and persistence involves strategic approaches. Some of these approaches include source reduction and preventative strategies such as green chemistry innovations, phasing out the volatile PFAS compounds from industrial processes, coatings, and surfactants. Advanced solutions such as catalytic filtration, supercritical water oxidation (SCWO), and engineered adsorbents show considerable promise but require further optimization, cost reduction, and demonstration at scale. Deploying sensor-based monitoring tools and fluorine-specific analyzers is essential for identifying and responding to emissions quickly. Investment in standardized air sampling protocols and detection methods is a key enabler. Regulatory frameworks also need to be updated to address airborne PFAS directly. Clear guidance on allowable limits, best available technologies (BAT), and emission thresholds will accelerate technology adoption.

Furthermore, insights from AI-enhanced modelling, techno-economic assessments, and life cycle analyses should inform not only the selection of control technologies but also long-term planning for facility design, emissions compliance, and public health protection. Continued investment in basic and applied research, especially on PFAS transformation chemistry, health risk assessments, public awareness and atmospheric fate, will be essential to closing current knowledge gaps.

References

1. Stockholm Convention on Persistent Organic Pollutants (POPs). [https://www.pops.int/Implementation/IndustrialPOPs/PFAS/Overview/tabid/5221/Default.aspx], (2022)

2. US EPA Fact Sheet PFOA & PFOS Drinking Water Health Advisories, [https://www.epa.gov/sites/default/files/2016-05/documents/drinkingwaterhealthadvisories_pfoa_pfos_5_19_16.final_.1.pdf], (2016)
3. USEPA. PFAS Strategic Roadmap: U.S. Environmental Protection Agency. https://www.epa.gov/pfas (2022)
4. G Codling, C. Halsall, L. Ahrens, S. D. Vento, K. Wiberg, M. Bergknut, H. Laudon, R. Ebinghaus, *The fate of per- and polyfluoroalkyl substances within a melting snowpack of a boreal forest*, Environmental Pollution, 191, 190-198, (2014)
5. A. Dreyer, I. Weinberg, C. Temme, R. Ebinghaus, *Polyfluorinated compounds in the atmosphere of the Atlantic and Southern Oceans: Evidence for a global distribution,* Environmental Science & Technology, 43(17), 6500–6507 (2009)
6. I. Ross, J. McDonough, J. Miles, P. Storch, P. T. Kochunarayanan, E. Kalve, J. Hurst, S. S. Dasgupta, J. Burdick, *A review of emerging technologies for remediation of PFAS*, Remediation Journal, 28(2), 101–126, (2018)
7. M. Ahmad, A. U. Rajapaksha, J Eun Lim 2, Ming Zhang 3, Nanthi Bolan 4, Dinesh Mohan 5, Meththika Vithanage 6, Sang Soo Lee 2, Yong Sik Ok 7, *Biochar as a sorbent for contaminant management in soil and water: A review*, Chemosphere, 99, 19–33, (2014)
8. M. Krause, E. Thoma, E. Sahle-Demessie, B. Crone, A. Whitehill, E. Shields, B. Gullett, *Supercritical water oxidation as an innovative technology for PFAS destruction*, Environmental Science and Technology, American Chemical Society, Washington, DC, 148(2):NA, (2022)
9. S.D. Currie, D.B. Benson, Z.R. Xie, J.S. Wang, L. Tang. Utilization of Artificial Intelligence Coupled with a High-Throughput, *High-Content Platform in the Exploration of Neurodevelopmental Toxicity of Individual and Combined PFAS,* J Xenobiot, (2025)

Vacuum–Suction Technology integrated with water, gas and solid waste treatment: A comprehensive solution for urban and industrial environments in Vietnam

Nguyen Duc Luong[1*], *Huynh Kim* Ho[2], *Huynh Thuan* Hieu[2] and *Nguyen Nhu* Y[1]

[1]Faculty of Natural Resources and Environment, Vietnam National University of Agriculture
[2]Hieu Thao Golden Company Limited, Gia Lam distrisct, Ha Noi city

Abstract. Water pollution, industrial exhaust gases, and hazardous solid waste pose significant challenges for Vietnam and other developing countries, while conventional treatment technologies often address only a single group of pollutants, with low efficiency and high operating costs. This study introduces the Vacuum–suction technology as an integrated solution capable of simultaneously treating wastewater, exhaust gases, and solid waste within a single closed system. The operating principle combines vacuum suction and compression pressure to maximize gas–liquid mixing and contact within the enclosed Octola reactor, thereby enhancing neutralization, precipitation, and absorption efficiency. Results show that, for industrial wastewater, the system achieved removal efficiencies of 87.1% for COD, 88.6% for TSS, 90.0% for NH_3–N, >85% for heavy metals, and 99.4% for total coliforms, meeting both Vietnamese QCVN and WHO standards for water reuse. For exhaust gases, the system removed 91.8% SO_2, 85.7% NOx, >95% HCl and HF, >90% VOCs and CO, and 92% total particulates, while reducing chemical consumption by 30–50% compared with conventional wet scrubbers. In solid waste treatment, leachable heavy metals from sludge and bottom ash decreased by more than 98%, with a 90% reduction in landfill volume and recovery of valuable metals. Compared with current technologies, this system reduces installation footprint by up to 40%, shortens hydraulic retention time to 2–4 hours, and enables resource recovery after treatment. With its high treatment efficiency, multi-functional integration, and broad application potential, Vacuum–suction technology offers a viable solution for waste management and environmental protection towards a circular economy.

1 Introduction

Water resource pollution, industrial emissions, and hazardous solid waste remain major challenges in Vietnam and other developing countries [1]. Conventional treatment technologies such as biofilters, bio-scrubbers, wet scrubbers, or ozone aeration typically

* Corresponding author: ndluong3399@gmail.com

© The Authors, published by EDP Sciences. This is an open access article distributed under the terms of the Creative Commons Attribution License 4.0 (https://creativecommons.org/licenses/by/4.0/).

address only a single group of pollutants, exhibiting low efficiency under real operating conditions and incurring high operational costs [2]. The Vacuum–suction technology is proposed as an integrated, multifunctional solution that enables natural oxygen supply to water sources while simultaneously treating exhaust gases, wastewater, and hazardous solid waste within a single system [3]. Its operating principle combines vacuum suction with compression pressure to convey and thoroughly mix gas and liquid streams inside a closed Octola reactor, thereby maximizing oxygen dissolution efficiency and enhancing gas neutralization [4]. Compared with conventional technologies, the system offers several advantages: integration of multiple functions in a single process line; comprehensive removal of diverse pollutants; reduced operating costs through the use of low-cost lime $Ca(OH)_2$; the ability to process moist waste; and recovery of resources after treatment [5]. With its flexibility and high performance, this technology is well-suited for implementation in urban areas, industrial parks, and craft villages, contributing to environmental ecosystem improvement, enhancing aquaculture productivity, and promoting green economic development [1], [6].

1.1 Vacuum Suction Technology

The Vacuum–suction technology operates on the principle of creating a pressure differential to convey and mix gas–liquid streams within a closed Octola reaction environment. Vacuum suction combined with compression pressure facilitates the capillary introduction of gas into the treatment solution, thereby maximizing the contact area and reaction efficiency. O_2 supply for water sources: Natural oxygen is drawn in and dispersed into ultra-fine bubbles through the water column, ensuring complete dissolution and minimizing gas loss compared to conventional forced-aeration methods [7]. Exhaust gas treatment: Hazardous gases are directed into an alkaline solution ($Ca(OH)_2$, NaOH, KOH, etc.), where they are continuously dispersed and neutralized to form neutral salts or insoluble precipitates [8]. Wastewater and solid waste treatment: Wastewater is pretreated to remove toxic gases using H_2O_2 or NaOCl before neutralization with alkali; solid waste is incinerated in a fluidized-bed furnace using natural O_2. The flue gas generated is then cooled and neutralized in a manner similar to that applied for industrial exhaust gas.

This principle enables multi-task integration within a single system, reducing the number of intermediate units, minimizing secondary emissions, and optimizing operating costs [9]

1.1.1 Application of Vacuum–Suction Technology Technology in Industrial Wastewater Treatment

The overall integrated treatment sequence of Vacuum–Suction Technology applied to industrial wastewater is illustrated in Fig.1, which presents the complete process flow diagram from degassing to neutralization, sedimentation, and oxygenation. Industrial wastewater contains numerous toxic compounds such as inorganic acids, strong bases, heavy metals, and volatile organic compounds (VOCs), all of which must be thoroughly treated before discharge. The Vacuum–suction technology applies a sequential integrated process: toxic gases are removed using H_2O_2 and NaOCl in the degassing Octola, with the released gases drawn under vacuum and neutralized in an alkaline solution; pH is adjusted and metals are precipitated using $Ca(OH)_2$ or NaOH/KOH in the neutralization Octola; solids are separated via zigzag sedimentation and vacuum-assisted removal of metal hydroxide sludge; finally, the water is softened with O_2 before being discharged or reused [10],[11].

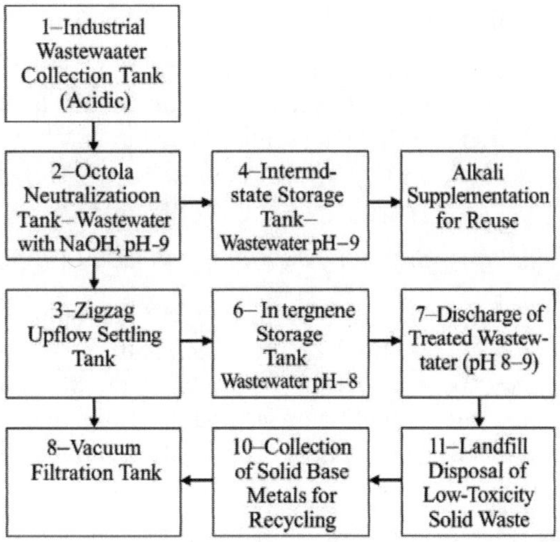

Fig.1. Process flow *diagram* of industrial wastewater treatment using Vacuum–Suction Technology technology

This method enables the simultaneous treatment of multiple types of pollutants within a closed-loop system, reducing installation footprint, optimizing operational costs, and allowing for the recovery of recyclable minerals.

1.1.2 Application of Vacuum–Suction Technology in Gas Treatment

Industrial exhaust gases often contain a mixture of pollutants such as SOx, NOx, CO, CO_2, HCl, HF, NH_3, VOCs, furans, dioxins, and metal oxide dust, generated from processes including combustion, metallurgy, cleaning, electroplating, and chemical manufacturing [12]. Without proper treatment, these gases contribute to air pollution, acid rain, and exert toxic effects on both living organisms and humans [13]. The Vacuum–suction technology treats exhaust gases by directing the gas stream through a neutralization Octola reactor operating under vacuum conditions combined with compression pressure. The exhaust is continuously dispersed into an alkaline solution ($Ca(OH)_2$, NaOH, KOH), where neutralization reactions convert acidic gases into neutral salts (e.g., $CaSO_4$, NaCl) and insoluble precipitates. Metal oxide dust is wetted, settled, and separated using a vacuum filtration tank [14]. The configuration and operational steps of the Vacuum–Suction Technology system for flue gas treatment are **shown in** Fig.2**,** illustrating the continuous gas dispersion, neutralization reactions, and particulate removal stages.

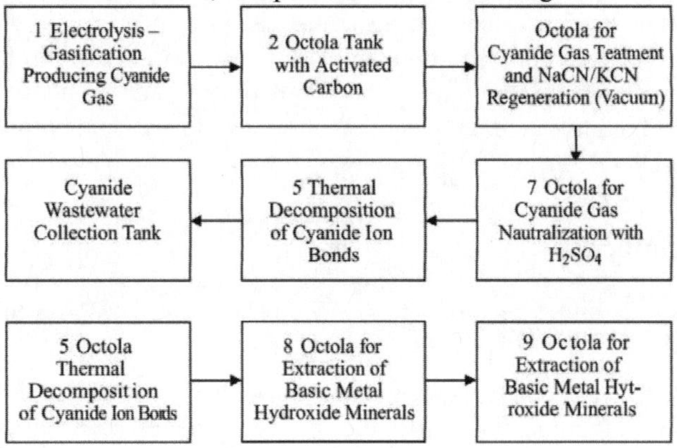

Fig.2. Flow Diagram of Flue Gas Treatment Using Vacuum–Suction Technology

This application enables the simultaneous removal of multiple pollutant components within a single device, achieving removal efficiencies exceeding 90% for most acidic gases and particulates, thereby reducing the need for multiple separate systems, saving installation space, and lowering operating costs [15].

1.1.3 Application of Vacuum–Suction Technology in Solid Waste Treatment

The complete process for solid waste treatment using Vacuum–Suction Technology is presented in Fig.3, including the fluidized-bed incineration, flue gas cooling, alkaline neutralization, and ash/metal recovery steps. Industrial and municipal solid waste, particularly that containing heavy metals, toxic organic compounds, or non-biodegradable components, can cause soil and groundwater contamination and release hazardous emissions if not properly treated [16].

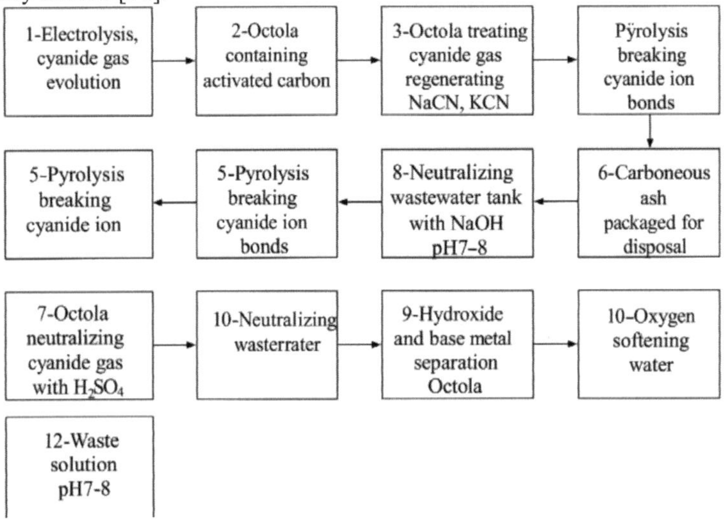

Fig.3. Flow Diagram of Solid Waste Treatment Using Vacuum–Suction Technology

The Vacuum–Suction Technology technology is applied to solid waste treatment through a multi-stage fluidized bed incinerator, utilizing combustion energy from natural O_2 in the air. The process maintains a temperature of 600–800 °C, ensuring complete decomposition of hazardous organic components without the need for fossil fuels [17]. The flue gas and steam generated are directed through a cooling system and subsequently neutralized in the Octola reactor with a $Ca(OH)_2$ solution, converting acidic gases into neutral salts and allowing metal oxide dust to settle in a vacuum filtration tank [18]. This method enables the incineration of moist waste without prior drying, reduces both the volume and toxicity of waste, conserves landfill space, and facilitates the recovery of bottom ash for reuse as construction materials or for extracting valuable metals [19].

1.2 Sampling Methods

1.2.1. Industrial wastewater

Wastewater samples were collected in accordance with TCVN 6663 (ISO 5667) and ISO 5667-10 guidelines at two locations: after the equalization tank (before treatment) and after the final treatment unit (after treatment). Time-composite samples (collected every 15–30 minutes over a 24-hour period) were used to analyze COD, TSS, NH_3–N, heavy metals (Pb, Zn, Cu), oil and grease, and coliform parameters. Grab samples were collected for rapid on-site measurement of pH and NH_3–N. Sample preservation followed APHA Standard Methods, including acidification with HNO_3 to pH < 2 for metal samples, refrigeration at 4 ± 2 °C, and analysis within the prescribed holding times.

1.2.2. Exhaust gas

Flue gas samples were collected using the isokinetic sampling method in compliance with EN 13284-1 (total particulate matter), US EPA Method 6 (SO_2), Method 7E (NOx), Method 26A (HCl, HF), and Method 25A (VOCs). Sampling points were located along straight duct sections meeting flow development requirements. Each sampling run lasted 60–120 minutes, with a minimum of 2–3 runs per source. Gas velocity, temperature, and flow rate were measured simultaneously to standardize emission load calculations

1.2.3. Solid waste, sludge – ash

Solid waste samples were collected and prepared as representative samples in accordance with ASTM D5231 and US EPA SW-846, Chapter 9. Samples were obtained from discharge conveyors, sludge pits, or ash hoppers, combining multiple random subsamples during the operating shift, then homogenized and quartered to produce a composite representative sample. Leaching tests were conducted following the Toxicity Characteristic Leaching Procedure (TCLP; US EPA Method 1311) to assess the potential release of metals and hazardous compounds.

1.2.4. Quality Assurance and Quality Control (QA/QC)

At least 10% of the total number of samples were allocated for QA/QC purposes, including field blanks, duplicate samples, and spiked samples. All analytical instruments were regularly calibrated, with acceptable relative percent difference (RPD) limits of ≤20% for wastewater and ≤30% for solid waste. The chain of custody (CoC) was maintained throughout the entire sampling and transportation process.

2 Research Results

2.1. Industrial Wastewater Treatment

Analysis of wastewater samples collected at two points (before and after treatment) demonstrated that the Vacuum–Suction Technology technology achieved high removal efficiencies for most pollution indicators (Figure 4). COD was reduced from 350 to 45 mg/L (87.1%), TSS from 220 to 25 mg/L (88.6%), and NH_3–N from 35 to 3.5 mg/L (90.0%). Heavy metal concentrations (Pb, Zn, Cu) decreased by more than 85%, oil and grease content dropped by 90%, and total coliform counts were reduced by 99.4%

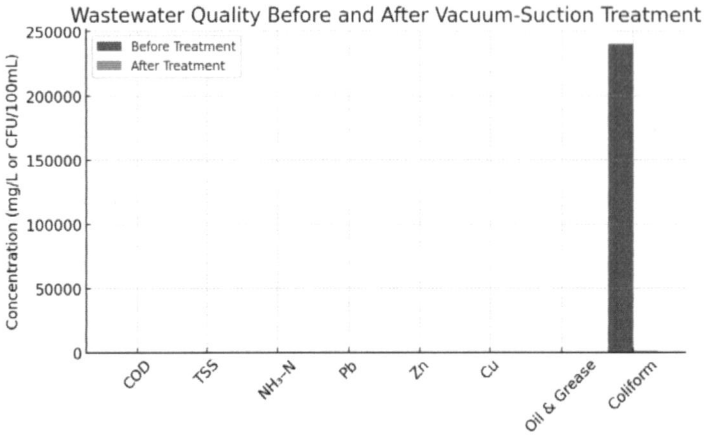

Fig.4. Industrial wastewater quality before and after treatment using Vacuum–suction technology

Compared with conventional constructed wetlands or sequencing batch reactor (SBR) systems, which typically require large installation areas and multiple sequential treatment stages [20], the Vacuum–Suction Technology technology integrates physico–chemical–biological processes within a single closed module. This configuration reduces the installation footprint by up to 40% and shortens the hydraulic retention time (HRT) to 2–4 hours, while still meeting current QCVN standards and WHO guidelines for water reuse [21].

2.2. Exhaust Gas Treatment

Analysis of flue gas samples before and after treatment showed a significant reduction in acidic gases, VOCs, and particulate matter (Figure 5). SO_2 concentrations decreased by 91.8%, NOx by 85.7%, HCl and HF by more than 95%, VOCs and CO by more than 90%, and total particulate matter by 92%.

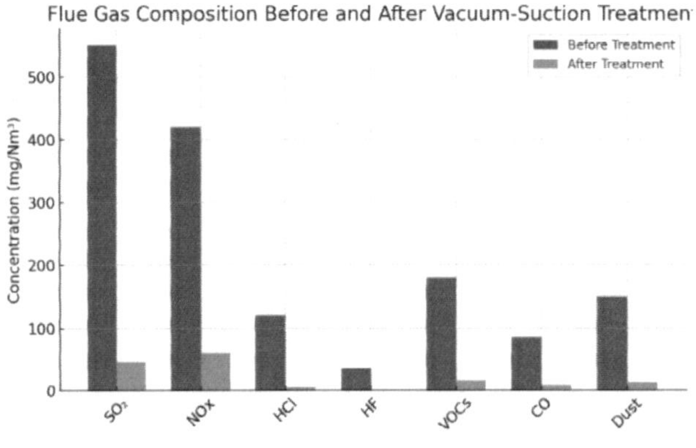

Fig.5. Exhaust gas quality before and after treatment using Vacuum–suction technology

Compared with conventional wet scrubbers, which require multiple absorption stages and are generally ineffective for VOC removal [22], the Vacuum–Suction Technology technology enables the simultaneous treatment of multiple pollutant components within a single unit. It reduces chemical consumption by 30–50% due to the enhanced gas absorption efficiency provided by the vacuum reaction environment. These results are comparable to or exceed those reported in international studies on advanced scrubber designs [23], particularly in the removal efficiency of HF and fine particulate matter.

2.3. Solid Waste Treatment

The analysis results indicate that the leachable metal content from sludge and bottom ash decreased by over 98% after treatment (Figure 6). Pb decreased from 3.5 to 0.05 mg/L, Zn from 8.0 to 0.12 mg/L, and Cu from 2.2 to 0.03 mg/L — all well below the limits specified by the US EPA and QCVN 07:2009/BTNMT.

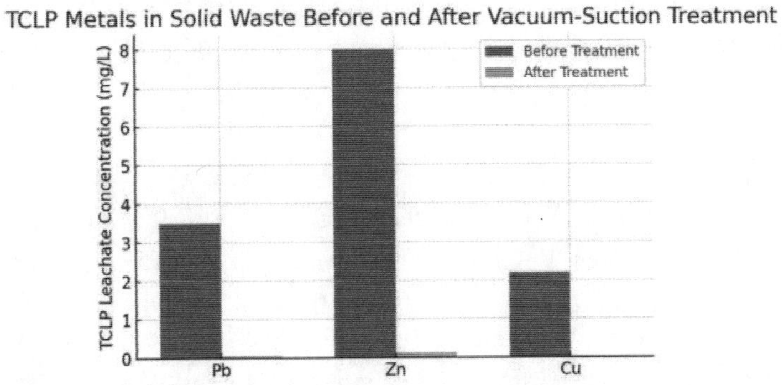

Fig.6. Heavy metal content in solid waste before and after treatment using Vacuum–suction technology

Compared with conventional solidification/stabilization methods using cement or lime, which typically increase waste volume by 10–20% and generate hydration heat [24], the Vacuum–Suction Technology technology simultaneously incinerates and stabilizes hazardous organics while separating and recovering valuable metals from bottom ash, thereby reducing landfill volume by up to 90%. This approach aligns with the "waste-to-resource" trend in modern solid waste management [25].

3 Conclusion

The Vacuum–Suction Technology technology has demonstrated outstanding performance in the integrated treatment of industrial wastewater, flue gas, and hazardous solid waste within a single closed-loop system, achieving high removal efficiencies and significantly lower operating costs compared with conventional technologies. Results show that the system achieved removal efficiencies of 87.1% for COD, 88.6% for TSS, 90.0% for NH_3–N, >85% for heavy metals, 90% for oil and grease, and 99.4% for total coliform, fully meeting both Vietnamese QCVN standards and WHO guidelines for water reuse, while reducing installation area by 40% and shortening the hydraulic retention time (HRT) to 2–4 hours compared with SBR or constructed wetlands. For flue gas treatment, the technology removed 91.8% of SO_2, 85.7% of NOx, >95% of HCl and HF, >90% of VOCs and CO, and 92% of total particulate matter, while reducing chemical consumption by 30–50% compared with conventional wet scrubbers due to the enhanced reaction efficiency in the vacuum environment. In solid waste treatment, leachable metal concentrations from sludge and bottom ash decreased by >98%, falling below US EPA and QCVN limits, while reducing landfill mass by up to 90% and enabling recovery of valuable metals and recyclable minerals. The capability to simultaneously treat multiple pollutant types, including high-moisture waste, combined with resource recovery potential and minimization of secondary emissions, demonstrates that Vacuum–Suction Technology technology is a suitable solution for deployment in industrial zones, urban areas, and craft villages, contributing to the realization of circular economy goals and sustainable development.

References

1. Q. Wang, Y. Wang, Z. Li, C. Zhang, L. Li, W. Xu, Water Sci. Technol. 85, 3066 (2022).
2. A. Bani-Melhem, M. Abbas, Desalination 273, 225 (2011).
3. S. M. Jang, J. H. Kim, Y. S. Lee, Sep. Purif. Technol. 193, 318 (2018).
4. A. P. Gitis, M. Rothenberg, Water Treatment: Advanced Processes and Technologies (World Scientific, Singapore, 2019).
5. J. Zhao, Q. Jiang, B. Xu, Chem. Eng. J. 352, 117 (2018).
6. T. S. Lee, H. M. Oh, D. W. Cho, J. Ind. Eng. Chem. 69, 285 (2019).
7. M. T. Nguyen, H. V. Tran, VNU J. Sci.: Earth Environ. Sci. 38, 45 (2022).
8. J. Yang, H. Chen, Y. Chen, Environ. Sci. Pollut. Res. 28, 24315 (2021).
9. US EPA, SW-846 Test Methods for Evaluating Solid Waste, Physical/Chemical Methods (United States Environmental Protection Agency, Washington D.C., 2020).
10. APHA, Standard Methods for the Examination of Water and Wastewater, 23rd edn. (American Public Health Association, Washington D.C., 2017).
11. ISO, ISO 5667-10:2020 – Water quality — Sampling — Part 10: Guidance on sampling of waste waters (International Organization for Standardization, Geneva, 2020).
12. EN 13284-1:2017 – Stationary source emissions — Determination of low range mass concentration of dust — Part 1: Manual gravimetric method (European Committee for Standardization, Brussels, 2017).
13. US EPA Method 6, Determination of Sulfur Dioxide Emissions from Stationary Sources (United States Environmental Protection Agency, Washington D.C., 2017).
14. US EPA Method 7E, Determination of Nitrogen Oxides Emissions from Stationary Sources (Instrumental Analyzer Procedure) (United States Environmental Protection Agency, Washington D.C., 2017).
15. US EPA Method 26A, Determination of Hydrogen Halide and Halogen Emissions from Stationary Sources (United States Environmental Protection Agency, Washington D.C., 2017).
16. US EPA Method 25A, Determination of Total Gaseous Organic Concentration Using a Flame Ionization Analyzer (United States Environmental Protection Agency, Washington D.C., 2017).
17. TCVN 6663 (ISO 5667) – Chất lượng nước – Lấy mẫu (Bộ Khoa học và Công nghệ, Hà Nội, 2011).
18. QCVN 07:2009/BTNMT – Quy chuẩn kỹ thuật quốc gia về ngưỡng chất thải nguy hại (Bộ Tài nguyên và Môi trường, Hà Nội, 2009).
19. World Health Organization, Guidelines for Drinking-water Quality, 4th edn., incorporating the 1st addendum (WHO, Geneva, 2017).
20. A. K. Verma, R. R. Dash, P. Bhunia, J. Water Process Eng. 1, 22 (2014).
21. Y. H. Song, S. J. Kim, Chemosphere 251, 126383 (2020).

22. R. G. Anthony, C. A. Samuelson, Ind. Eng. Chem. Res. 28, 1036 (1989).
23. Y. F. Cheng, J. M. Smith, Environ. Technol. 27, 123 (2006).
24. H. A. van der Sloot, P. Kosson, F. Sanchez, Cement-based stabilization/solidification: An old technology with new applications (RILEM Publications, Paris, 2017).
25. J. P. Gourdon, Waste Manage. 33, 1395 (2013).

t pp s/dd o.gpr5dl 0gl 0(1de3sconfd606)) 900100(

Integrating Indigenous Governance into Nature-Based Solutions for Climate and Biodiversity Resilience

Rismawati Nur[1*], *Retno* Meilani[1], *Alfin* Dwi Novemyanto[1]

[1]Faculty of Law, Universitas Gadjah Mada, Yogyakarta, Indonesia

Abstract. The accelerating climate crisis underscores the limitations of state-centric and technocratic approaches to environmental governance. Although Nature-Based Solutions (NbS) are increasingly promoted as strategies for climate adaptation and biodiversity conservation, prevailing frameworks often neglect Indigenous governance systems that have long sustained ecosystems through customary law, ecological knowledge, and cultural values. This article positions Indigenous governance as a pivotal dimension of NbS, emphasizing its capacity to integrate ecological stewardship with social justice and intergenerational equity. Drawing on the case of the Ammatoa Kajang community in South Sulawesi, Indonesia, the study illustrates how Indigenous forest classifications and customary norms safeguard ecological balance while reinforcing cultural resilience. Employing a mixed-methods approach, combining ethnography, geospatial analysis, and reflective narrative. The research demonstrates that Indigenous-led governance provides legitimacy and inclusivity frequently absent in state-driven conservation initiatives. The findings highlight the importance of legal pluralism and co-management models that recognize Indigenous rights, thereby advancing NbS that are ecologically robust, socially just, and culturally sustainable.

1 Introduction

Climate change is one of the most serious challenges facing humanity in this century. Scientific evidence consistently shows a significant increase in global temperatures over time. The latest data shows that in 2024, the global land average annual temperature anomaly reached more than 2.0 °C compared to the normal period of 1951–1980, making it the highest record since temperature records began in 1880 [1]. Figure 1 shows the trend in annual and seasonal temperature changes, which have increased almost linearly since the mid-1980s.

* Corresponding author : rismawatinur@mail.ugm.ac.id

© The Authors, published by EDP Sciences. This is an open access article distributed under the terms of the Creative Commons Attribution License 4.0 (https://creativecommons.org/licenses/by/4.0/).

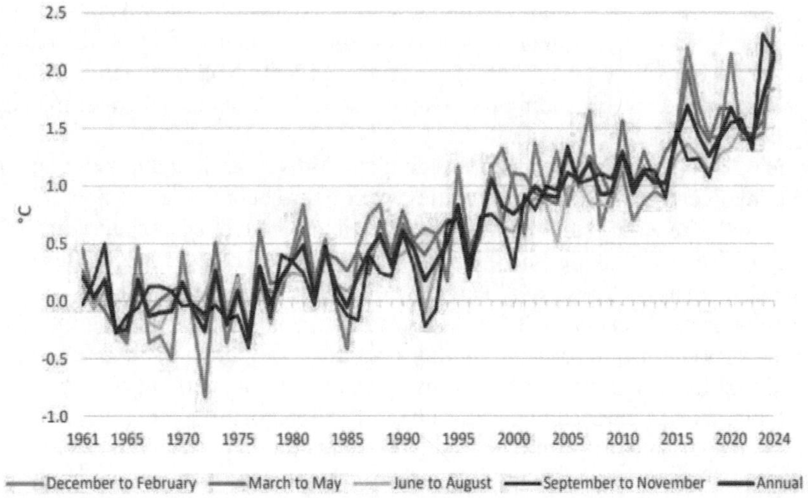

Fig. 1. Mean annual and seasonal temperature change on land

The data shows that the last ten years have been the hottest period in history, with an average increase of around 1.5 °C. This threshold is a critical point set in the *Paris Agreement* to prevent catastrophic climate impacts. Several seasons have recorded temperature anomalies above 2.0 °C, indicating an accelerating climate crisis with direct impacts on ecosystems and human life. Figure 2 shows that while almost no population lived in countries exceeding 1.5 °C before 1980, this figure has risen sharply since the early 2000s, reaching more than 60% of the world's population by 2024. This data reinforces the fact that climate change is not a future threat, but a reality that is now being felt by most of the global community.

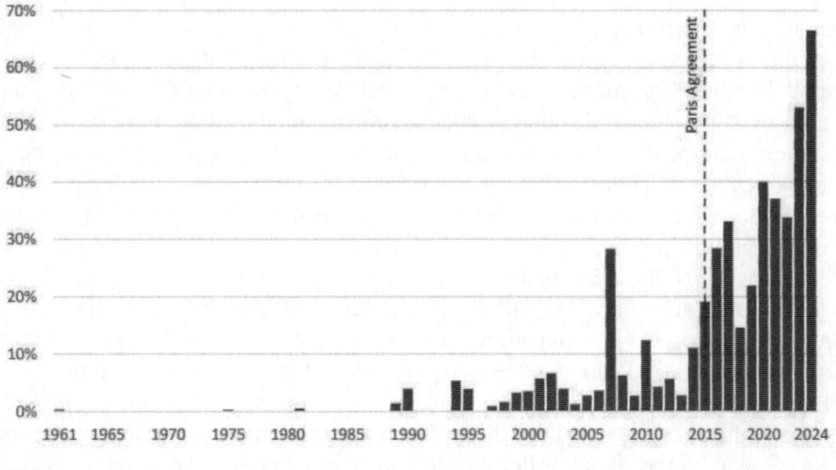

Fig. 2. Share of population living in countries with warming above 1.5 °C in world population

Floods, droughts, storms, and shifts in extreme weather patterns are occurring more frequently and with greater intensity. These changes increase socio-economic vulnerability, threaten food security, and increase the risk of livelihood loss, especially for vulnerable communities such as indigenous peoples who depend directly on nature for their livelihoods.

Thus, climate impacts must be viewed not only as ecological problems but as multidimensional challenges requiring integrated solutions. In response, sustainability-oriented approaches are increasingly emphasized in global climate policy, with Nature-based Solutions (NbS) emerging as a prominent strategy.[2-3]. NbS is characterized by its ecosystem-based approach, including biodiversity protection, improved water management, and strengthened social cohesion [4].

The success of NbS is highly dependent on a supportive social context and governance. Without the involvement of local communities, NbS risks becoming a technocratic program that ignores the needs of communities in the field. Active community participation, transparency in decision-making, and the principle of justice are key prerequisites for NbS to be truly inclusive. Thus, NbS should be understood not only as an ecological strategy, but also as a social process that requires cross-actor collaboration. In this context, indigenous peoples have a very strategic role to play. For centuries, they have developed management systems based on local wisdom that have proven capable of maintaining ecological balance. Previous studies have demonstrated the ability of indigenous peoples to adapt to socio-ecological changes. The participation of indigenous peoples in conservation has proven to promote a more effective and equitable ecological restoration process.

These dynamics indicate that customary governance should be positioned on an equal footing with modern scientific approaches. International recognition of Indigenous peoples is reflected in legal instruments such as the UN Declaration on the Rights of Indigenous Peoples (UNDRIP), which affirms their rights to land, territories, and resources, as well as their participation in decision-making processes. Moreover, Local and Indigenous Knowledge Systems (LINKS) provide important insights for ecosystem management and restoration. Through their longstanding relationship with forests and land, Indigenous communities have developed tradition-based management mechanisms that remain effective in sustaining ecological balance [5]. Indigenous peoples' contributions remain largely overlooked in modern conservation. Many conservation areas, including national parks and biosphere reserves, restrict their access to ancestral lands, despite evidence that Indigenous communities have long maintained ecosystem sustainability. These contradictions create social tensions and weaken conservation outcomes. Research also shows that Indigenous peoples continue to organize and defend their rights against external development pressures that threaten their territories. [6].

In addition to recognition issues, the implementation of NbS also faces technical and structural obstacles. Various case studies show that the application of NbS is often hampered by limited funding, weak institutional capacity, and a lack of policy support [2]. Unplanned urbanization and uncontrolled land exploitation further degrade ecosystems, undermining NbS implementation. This confirms that NbS should not be viewed solely as a technical approach but rather requires participatory governance transformation. Within this framework, indigenous community governance can be understood as a concrete manifestation of NbS. This system not only ensures the sustainability of resource management but also brings about social justice and cultural resilience. By placing indigenous knowledge as its foundation, NbS has the potential to produce more inclusive and sustainable solutions as a climate change mitigation effort that does not sacrifice the rights of communities.

Based on the above description, this paper will explain that indigenous governance is a crucial element of nature-based solutions for climate mitigation and ecosystem resilience. By examining the relationship between indigenous governance, NbS, and global policies, this article offers a comprehensive perspective on how local knowledge can be integrated into climate strategies. Specifically, this study uses the Kajang indigenous community in South Sulawesi, Indonesia, as a case study to explore how traditional governance systems contribute to ecosystem conservation and climate resilience. This approach is expected to encourage synergy between modern science, international policy, and local knowledge systems to produce effective and equitable climate solutions.

2 Research Method

This study uses a mixed-method approach by integrating three complementary methodological components. First, ethnographic observations were conducted in conjunction with in-depth interviews with key informants, particularly traditional leaders, community members, and relevant policy makers, to gain a contextual understanding of indigenous community governance. Second, this study utilizes qualitative analysis of geospatial data, including maps of indigenous territories and changes in land cover, to assess the role of indigenous-based governance in preserving ecosystems. Third, a reflective approach is used through first-person narratives [7], which reconstructs the researcher's empirical experiences and enriches the interpretation of the interaction between customary norms and *Nature-based Solutions* strategies. The integration of these three methods allows for a more comprehensive analysis of the potential of indigenous community governance as a nature-based solution in climate change mitigation and ecosystem resilience strengthening.

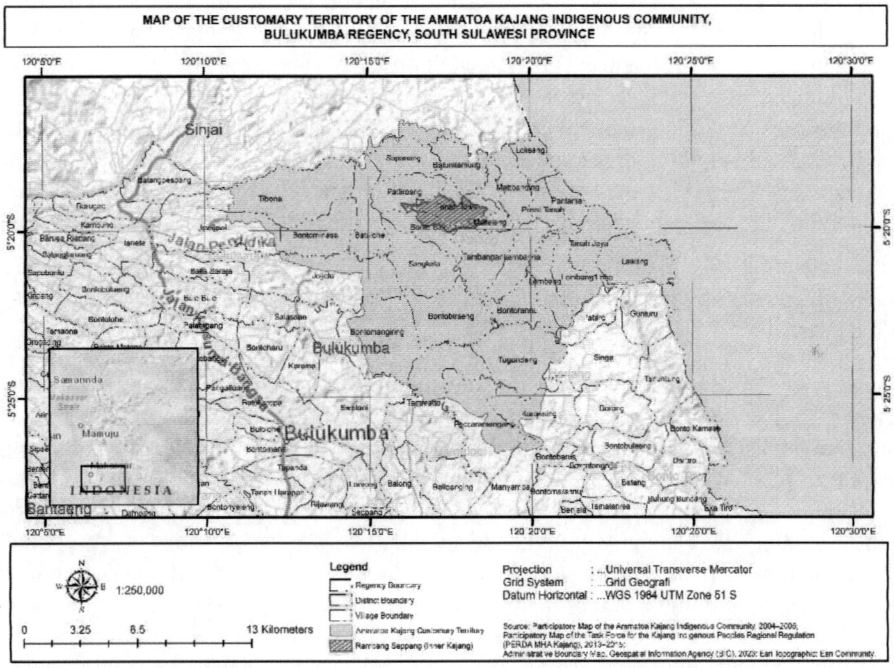

Fig. 3. Map of the Customary Territory of the Ammatoa Kajang Indigenous Community, Bulukumba Regency, South Sulawesi Province.

This study examines the governance of customary forests within the Ammatoa Kajang Indigenous Community in Bulukumba Regency, South Sulawesi. The community occupies a traditionally recognized territory with clearly defined boundaries, as illustrated in Fig. 3, which provides essential spatial context for understanding the jurisdictional framework that shapes land-use governance and customary authority. The analysis integrates data from ethnographic observations and in-depth interviews with traditional leaders, community members, and policymakers, complemented by spatial analysis using maps of customary areas and land-cover change. These findings are further reinforced by the researcher's narrative reflections to provide a holistic understanding of how customary norms interact with Nature-based Solutions strategies.

3 Discussion

3.1 Indigenous-led Governance as a Pathway to Biodiversity Conservation

At the global level, Indigenous peoples are increasingly recognized for their effective management of the environment and natural resources, as well as their significant contribution to biodiversity conservation. Although they comprise only about 5% of the world's population, they help protect nearly 80% of global biodiversity, while their customary territories account for roughly 24% of global land and include most of the world's remaining intact forests. These figures indicate that Indigenous-led environmental governance goes beyond ecosystem preservation, integrating cultural, spiritual, and social values into conservation practices. From this holistic perspective, conservation is understood not only as an ecological effort but also as part of collective identity and intergenerational responsibility [8–10].

Indigenous knowledge-based environmental management has great potential to create an ethical space, which is a space for dialogue that allows for the equalization of traditional knowledge and modern science. This interaction not only broadens the perspective of conservation, but also produces more inclusive, fair, and sustainable strategies in responding to contemporary ecological challenges [10]. Therefore, customary governance should not be viewed merely as an alternative model, but as the main foundation for efforts to preserve biodiversity and strengthen ecosystem sustainability, whether at the local, national, or global level.

3.2 Indigenous Governance in Climate Change Mitigation and Ecosystem Resilience

Indigenous governance plays a strategic role in climate change mitigation and ecosystem resilience. Indigenous forests worldwide, including those in the Amazon, function as major carbon sinks and help maintain ecosystem balance. Global estimates show that Indigenous peoples and local communities manage about 24% of all above-ground carbon stored in tropical forests, amounting to more than 54 billion metric tons of carbon [11]. This confirms that the continuity of customary governance not only has implications for the preservation of the local environment but is also a key factor in achieving global carbon emission reduction targets. Without recognition and protection of the management rights of indigenous peoples, climate mitigation efforts risk losing their most important ecological foundation.

Traditional ecological knowledge passed down across generations reinforces this contribution. The practice of cultural burning in North America, for example, has been shown to reduce the risk of large fires while increasing ecosystem diversity. Similarly, indigenous agroforestry systems in Asia not only maintain soil productivity and food sources but also play a role in carbon sequestration and reducing vulnerability to climate change. With a value-based approach, indigenous peoples view ecosystems as an integral part of their social and spiritual lives, not merely as economic resources. The recognition of rights, collective responsibilities, and ecological wisdom inherent in indigenous societies are fundamental factors that make Nature-based Solutions (NbS) more effective because they strengthen socio-ecological resilience while promoting environmental justice.

3.3 Towards Inclusive Forest Management by the Kajang Indigenous Community in Indonesia

Forest management by the indigenous Kajang tribe of South Sulawesi demonstrates the existence of inclusive forest management by indigenous peoples. The system of classifying forests into various levels, namely *Borong Karamaka*, *Borong Batasayya*, and *Borong Luarayya*, not only describes the community's concrete practices in distinguishing between sacred, communal, and subsistence functions, but also demonstrates a structured ecological rationality. This can be seen in the following Table 1.

Table 1. Classification of the Ammatoa Kajang Indigenous Forest Areas

Types of Forests	Function
Borong Karamaka	*Borong Karamaka* is a forest area designated as a habitat for the flora and fauna of the customary area, so it cannot be touched by anyone for human activities. This forest is the place where *Tu Rie A'ra'na* (the indigenous people's belief in God) descends, so it is highly sacred. The only activities that can be carried out in the *Borong Karamaka* area are traditional rituals.
Borong Batasayya	This area is designated as a forest area, the proceeds of which are used to develop public facilities if timber is available and with the permission of Ammatoa as the traditional leader.
Borong Luarayya	This area is a forest area that can be managed by indigenous peoples to meet their daily needs, but such management must be based on the contents of the *Pasang Ri Kajang*, considering the balance of nature.

Source: author, 2025

As shown in Table 1, the institutional arrangement preserves a balance between conservation objectives and subsistence needs by embedding local practices within broader strategies for biodiversity protection and climate change mitigation. Governance inclusivity emerges through collective compliance with customary rules, the authority exercised by the Ammatoa, and the shared communal responsibility for safeguarding forest resources. Customary sanctions and cultural norms transmit ecological values across generations, illustrating that sustainability is sustained not only by formal regulatory frameworks but also by robust socio-cultural institutions. These dynamics foster a degree of social legitimacy that state regulatory mechanisms alone often fail to achieve. Furthermore, the Kajang community applies a mandatory regeneration principle that requires every harvested tree to be replaced and obliges individuals to plant trees prior to cutting, thereby ensuring continuous forest renewal and long-term ecological stability.

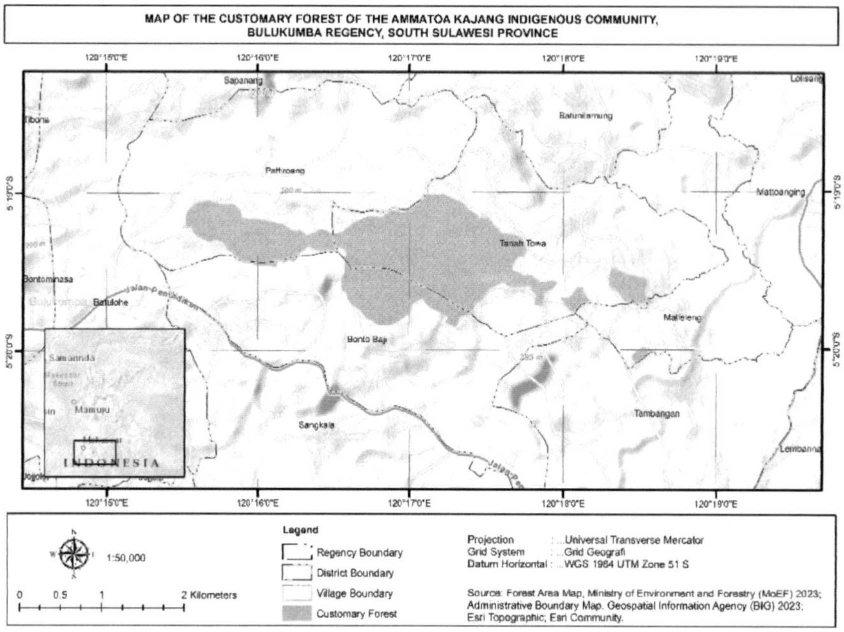

Fig. 4. Spatial distribution of the Ammatoa Kajang customary forest area.

The spatial distribution and ecological configuration of the Ammatoa Kajang customary forest, as illustrated in Fig. 4, provides a clear visual representation of the forest zones that underpin the community's governance system and resource-use regulations. Forest management by the Ammatoa Kajang Indigenous Community emphasizes the importance of the principle of environmental justice. With recognition of land and resource rights, communities can maintain authority in decision-making. This reality shows that customary governance is not subordinate but can coexist with and even complement national conservation policies. The Kajang Indigenous People provide a concrete example of a co-management model in which customary authorities, the state, and civil society can interact on an equal footing.

The experience of the Kajang indigenous community proves that inclusive customary governance can strengthen Nature-Based Solutions (NbS). By embedding ecological management into social institutions and cultural identity, this form of governance not only contributes to biodiversity conservation and carbon sequestration but also strengthens ecosystem resilience while preserving cultural sustainability. These findings critique and enrich the technocratic tendencies in NbS practices that often neglect social and cultural dimensions. Therefore, inclusivity in the case of the Kajang tribe is not only interpreted as participatory involvement, but also as the recognition and institutionalization of customary leadership as the center of environmental governance.

This study shows that customary ecological governance is not merely an alternative, but rather the main foundation for Nature-Based Solutions. While the NbS framework promoted by various international organizations often emphasizes technical interventions, such as reforestation or ecosystem restoration, customary governance emphasizes the social, cultural, and spiritual dimensions of ecological management. By embedding ecological responsibility into customary law and collective identity, this system can overcome the main weaknesses of state governance, namely low legitimacy and limited land ownership by communities.

One important insight from the case of the Kajang Indigenous People is the capacity of customary governance to integrate ecological resilience with cultural sustainability. This dual role is often overlooked in the dominant NbS discourse. The Kajang Indigenous Community shows that preserving ecosystems cannot be separated from protecting the rights and cultural integrity of indigenous peoples. Thus, customary governance promotes not only ecological outcomes but also social justice and intergenerational justice.

Fig. 5. (1-3)-1. Satellite map showing the Indigenous Regions of Ammatoa Kajang in Bulukumba, South Sulawesi; 2. Members of the Ammatoa Kajang community in traditional black attire during daily activities; 3. Traditional Ammatoa Kajang settlements surrounded by forest.

The spatial context and everyday interaction between the Kajang community and their forest environment, as illustrated in Fig. 5, provide an empirical foundation for understanding how customary norms shape environmental governance. For the indigenous Kajang tribe, rules regarding the division of forest areas are regulated in the Pasang ri Kajang. One of the tribal precept's states, "*Anjo natahang ri boronga karena pasang. Rettopi tanayya rettoi*," which means that the preservation of the forest is guaranteed because of tribal protection. If the forest is damaged, then the tribal order will also be disrupted. The Kajang tribe clearly demonstrates that humans, customs, and nature are inseparable entities. If nature is degraded, the survival of the community and the integrity of customary norms are likewise threatened. Conversely, if the rules are broken, nature will be destroyed, which will also have an impact on human survival.

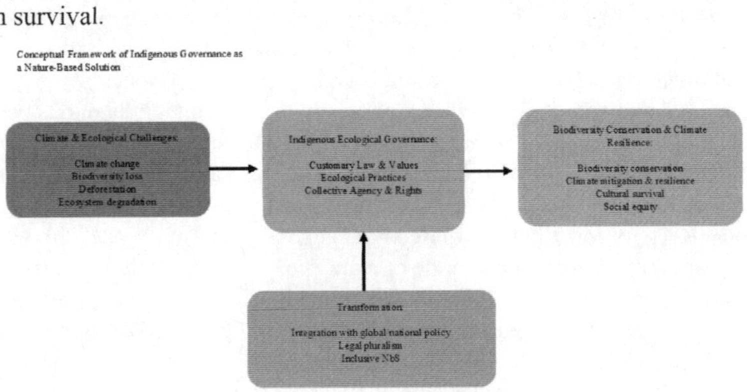

Fig. 6. Conceptual Framework of Indigenous Governance as a Nature-Based Solution

Figure 6 presents a conceptual framework that positions customary governance as a core component of NbS and linking ecological challenges with social outcomes. Processes of change, such as legal pluralism, the incorporation of customary rules into policy, and the recognition of indigenous peoples' rights, strengthen governance systems. This ultimately results in biodiversity conservation, reduced climate change impacts, and ecosystem resilience. Thus, this framework shows a recurring and mutually reinforcing relationship between governance, ecological sustainability, and cultural continuity. However, challenges remain. National development agendas and the expansion of extractive industries often reduce the living space of indigenous peoples, thereby weakening ecological and cultural resilience. Policy frameworks in Indonesia and other countries tend to place customary governance under positive state law. Without recognition of legal pluralism, the transformative potential of customary governance as a NbS will remain limited.

A comparative perspective reinforces the global implications of these findings. Research in the Amazon shows that legally recognized indigenous territories have lower rates of deforestation than surrounding areas. Similarly, Maori ecological governance in Aotearoa New Zealand has successfully integrated cultural values into the country's legal framework and offers a model for co-management. These cases demonstrate that incorporating indigenous governance into climate policy, both nationally and globally, is not only important but necessary to ensure the effectiveness of NbS.

4 Conclusions and Recommendations

This study concludes that indigenous peoples' ecological governance is not merely an alternative, but rather the main foundation of Nature-Based Solutions (NbS) in climate change mitigation, biodiversity conservation, and ecosystem resilience strengthening. The experience of the Kajang indigenous community in South Sulawesi shows how customary law, cultural values, and ecological knowledge are integrated into daily practices that ensure environmental sustainability and intergenerational justice. Unlike state-centric or technocratic approaches, customary governance embeds ecological responsibility into social and cultural institutions, giving it stronger legitimacy, inclusiveness, and sustainability. However, recognition of customary governance in national and international policy frameworks remains limited, often placed under state law and development agendas. Therefore, strengthening indigenous peoples' ecological governance is key to achieving biodiversity conservation that is fair, sustainable, and resilient to climate change.

Based on these findings, several recommendations can be proposed:

1) Recognition and Integration of Customary Governance in Policy
   Governments need to provide strong legal recognition of indigenous peoples' rights to land, territories, and resources, and integrate indigenous ecological knowledge into national and global strategies related to climate and biodiversity.
2) Strengthening Co-Management Models and Community Capacity
   An inclusive co-management framework between the state, indigenous peoples, and civil society is needed, accompanied by funding support and capacity-building programs to strengthen indigenous institutions and ecological management practices.
3) Global Recognition and Knowledge Exchange
   International organizations and climate governance regimes must prioritize indigenous governance in global policy, while facilitating knowledge exchange among indigenous communities worldwide to strengthen ecosystem resilience and environmental justice.

Thus, indigenous peoples' ecological governance can be positioned as a transformative pathway, so that biodiversity conservation and climate change mitigation are not only based on ecological aspects, but also on social justice and cultural sustainability.

The authors would like to acknowledge the Lembaga Pengelola Dana Pendidikan (LPDP), Ministry of Finance of the Republic of Indonesia, for its financial support of this publication. The authors also gratefully acknowledge Engki Fatiawan for his assistance in preparing the spatial data and maps used in this study.

References

1. F. FAO, *Temperature change statistics 1961–2024 – Global, regional and country trends*, **101** (Rome : FAOSTAT, 2025)
2. C. J. Effiong, "Climate justice in land use planning : Exploring the potential and challenges of nature-based solutions integration in Nigeria," *Journal of Environmental Management*, **377** (2025)
3. N. Kabisch, H. Korn, J. Stadler, and A. Bonn, Eds., *Nature-Based Solutions to Climate Change Adaptation in Urban Areas : Linkages between Science, Policy and Practice*. in Theory and Practice of Urban Sustainability Transitions. (Cham : Springer International Publishing, 2017)
4. S. E. Debele *et al.*, "Nature-based solutions can help reduce the impact of natural hazards : A global analysis of NBS case studies," *Science of The Total Environment*, **902** (2023)
5. G. Nimachow, R. Joshi, and O. Dai, "Role of indigenous knowledge system in conservation of forest resources–A case study of the Aka tribes of Arunachal Pradesh," *Journal of Traditional Knowledge*, **10** (2011)
6. D. C. Bahamonde, A. Casellas, and F. Araos, "Getting our sea back : Indigenous governance and biocultural conservation of coastal and marine commons," *Marine Policy*, **178** (2025)
7. L. MacNell, "A geo-ethnographic analysis of low-income rural and urban women's food shopping behaviors," *Appetite*, **128** (2018)
8. C. Sobrevila, *The Role of Indigenous Peoples in Biodiversity Conservation : The Natural but Often Forgotten Partners*. (Washington, D.C: The World Bank, 2008)
9. S. T. Garnett, N. D. Burgess, and I. Leiper, "A spatial overview of the global importance of Indigenous lands for conservation," *Nature Sustainability*, **369–374** (2018).
10. B. Vogel, L. Yumagulova, G. McBean, and K. A. C. Norris, "Indigenous-Led Nature-Based Solutions for the Climate Crisis : Insights from Canada," *Sustainability*, **14** (2022)
11. IWGIA, "The Indigenous World 2020: Editorial," iwgia.org. Accessed: Sep. 01, (2025)

 https://doi.org/(0.(05(/e3sconf/10166900100(

Integrating Waste Valorization and Green Energy: Biodiesel Production Using Microwave-Derived Chitosan Catalyst from Raw Tilapia (*Oreochromis Niloticus*) Scales

Tristan Roy Panaligan[1*], *Andrea Marie* Alam[1], *Jean Margaret* Berroya[1], and *Princess Lynelle Carysse* Torres[1]

[1]Mapúa Institute of Technology at Laguna, Mapúa Malayan Colleges Laguna, Cabuyao, Laguna, Philippines, 4025

Abstract. This study presents an integrated approach to biodiesel production using waste vegetable oil (WVO) and chitosan derived from raw tilapia (Oreochromis niloticus) scales, synthesized via microwave-assisted methods. The chitosan acted as a heterogeneous catalyst during the transesterification of pre-treated WVO. Key process parameters included esterification and transesterification stages, with acid value and volume changes monitored throughout. Characterization of the resulting biodiesel was conducted in accordance with the Philippine National Standard PNS/DOE QS 002:2015, focusing on total sulfur content, water and sediments, and Fatty Acid Methyl Ester (FAME) content. The final product exhibited compliant sulfur and water levels, though the FAME content remained below specification, attributed to limited glycerol separation time. Despite this, the process achieved a 64% conversion and reduced the acid value from 2.11 to 0.262 mg KOH/g, indicating high potential for further optimization. This research supports sustainable fuel development while promoting circular utilization of agrowaste in developing economies.

1 Introduction

The global rise in carbon dioxide (CO_2) emissions has contributed to an approximate 1 °C increase in global temperatures since the industrial era, with countries like the Philippines ranking as the 69th most polluted globally in 2022 [1]. A major contributor to greenhouse gas emissions, the transportation sector is expected to surpass coal-fired power plants in CO_2 emissions by 2040 at an annual growth rate of 5.8% [2]. The pressing need to cut CO_2 emissions has made the hunt for alternative renewable energy sources a top priority for countries like the Philippines that rely largely on fossil fuels. Made from plant or animal fats, biodiesel is a sustainable substitute for traditional diesel because it is non-toxic, biodegradable, and emits fewer exhaust pollutants, making it a perfect fit for the transportation industry [3].

[*] Corresponding author: trlpanaligan@mcl.edu.ph

© The Authors, published by EDP Sciences. This is an open access article distributed under the terms of the Creative Commons Attribution License 4.0 (https://creativecommons.org/licenses/by/4.0/).

In alignment with the Philippine government's Republic Act No. 9367 (Biofuels Act of 2006), this study highlights the abundant yet underexploited potential of waste vegetable oil (WVO) as a renewable, low-cost feedstock for biodiesel production. Even within the European Union—despite its progressive stance on waste management—substantial collection inefficiencies remain. It is estimated that the EU generates approximately 4 million tonnes of waste cooking oil (WCO) annually, based on a per capita production of 8 liters for a population of around 500 million. However, current collection efforts capture only about one-fourth of this potential volume, highlighting a significant gap between generation and recovery [4]. Additionally, Pakistan generated approximately 468,842 tonnes of WVO in 2023, reinforcing the relevance of national-scale valorization frameworks [5].

Despite global availability, the Philippines has yet to fully realize the potential of its own WVO stream. While comprehensive national data remain scarce, foreign industry trends and domestic policy statements suggest significant uncollected volumes from food establishments and households.

Beyond underutilization, improper WVO disposal presents a growing environmental concern. As highlighted in recent studies, the absence of structured WVO management contributes to serious pollution issues. For example, Hamdi et al. [6] reported that poor handling practices in restaurants, hotels, and food-processing industries—particularly those using retort technology—can lead to odor problems and oil film formation on water surfaces, thereby increasing biological oxygen demand and degrading aquatic ecosystems.

Given that WVO typically contains high levels of free fatty acids—ranging from 5% to 20%—pretreatment is essential to minimize soap formation and improve biodiesel yield [7]. Utilizing WVO not only reduces reliance on fossil fuels but also addresses environmental concerns associated with improper disposal, which can lead to pollution and public health risks [8], [9].

This study responds to these challenges by employing chitosan synthesized from tilapia fish scales as a biodegradable, heterogeneous catalyst—a novel approach that enhances sustainability. The chitosan was prepared via microwave-assisted demineralization, deproteinization, and deacetylation, as previously established in our earlier work [10], which enabled a more rapid and eco-efficient synthesis process. The same microwave technique was also applied to the transesterification step. Despite a limited 9-hour separation period for biodiesel and glycerol—shorter than typical reaction durations—the process demonstrated favorable energy performance and practical viability.

By valorizing both waste vegetable oil and fish-processing by-products, this work addresses critical environmental and energy issues while contributing to the United Nations Sustainable Development Goals (SDGs) 7 and 13.

2 Methodology

2.1 Chitosan Preparation as Heterogeneous Catalyst

Chitosan used in this study was synthesized from raw tilapia *(Oreochromis niloticus)* fish scales via microwave-assisted demineralization, deproteinization, and deacetylation, following the optimized procedure established in our previous work [10]. The prepared chitosan was subsequently employed as a heterogeneous catalyst in the transesterification of waste vegetable oil for biodiesel production.

 https://doi.org/(0.(05(/e3sconf/10166900100(

2.2 Preparation of WVO for Biodiesel Production

WVO was collected from a local restaurant in Muntinlupa, Philippines. The preparation procedure was adapted from Sahar et al. [2] with necessary modifications. Initially, the WVO was filtered using a 200-mesh sieve to eliminate coarse particulates and solid impurities. Following filtration, the oil was heated to 105 °C and maintained at boiling for 30 minutes to remove residual moisture. Once moisture removal was completed, the oil was allowed to cool to room temperature in preparation for the subsequent citric acid degumming process.

2.3 Citric Acid Degumming

Citric acid degumming was utilized to remove phospholipids and other polar impurities from the filtered oil, a critical step in improving biodiesel quality. Following the procedure described by El Knidri et al. [11] with slight modifications, 200 mL of the pre-treated oil was mixed with 6 mL of 30% (v/v) citric acid solution and heated to 70 °C under continuous stirring for 20 minutes. The mixture was subsequently cooled to 40 °C, at which point 2 mL of distilled water was added to facilitate phase separation. The emulsion was then centrifuged at 4000 rpm for 45 minutes, allowing the separation of the degummed oil from the gum residue.

2.4 Esterification

In order to reduce the amount of free fatty acids (FFAs) in the oil and improve the effectiveness of the transesterification procedure that followed, the esterification phase was used to transform the FFAs into methyl esters. This pretreatment is essential for avoiding soap production, which can have a negative impact on the yield and purification of biodiesel. Following the methodology outlined by Supraja et al. [12], with appropriate modifications, 40 mL of degummed waste vegetable oil was combined with 240 mL of methanol and 0.4 mL of sulfuric acid in a 500 mL round-bottom flask. The mixture was subjected to microwave irradiation for 1 minute, then transferred into a separatory funnel and allowed to settle under gravity for 45 minutes. This process produced two layers: the upper ester-rich phase and a lower aqueous layer. The esterified oil was carefully decanted, washed with warm distilled water to remove residual acid and methanol, and subsequently dried using sodium sulfate to ensure removal of moisture and impurities.

2.5 Transesterification

Transesterification was conducted to convert pretreated waste vegetable oil into biodiesel using chitosan as a heterogeneous catalyst. In a 500 mL round-bottom flask, 40 mL of pre-treated waste vegetable oil was combined with 240 mL of methanol and 0.4 g of chitosan derived from tilapia scales. The reaction mixture was subjected to microwave irradiation for 30 minutes with a ramping time of 4 minutes, following the procedure described by Babel et al. [7].

Upon completion of the reaction, the mixture was transferred to a separatory funnel and allowed to stand undisturbed to facilitate gravity separation of the two phases. The bottom glycerol-rich aqueous layer was drained, leaving the ester-rich oil layer. The ester oil was then washed with warm distilled water at approximately 55°C and gently agitated. After standing for 9 hours, the two layers were again separated, and the aqueous phase was removed, as described by Arachchige et al. [13].

To remove residual water, a dry washing step was performed by adding 4 to 5 g of sodium sulfate to the ester layer. The mixture was left to stand until the sodium sulfate clumped with moisture and impurities. The biodiesel product was then filtered to remove the desiccant and yield a purified biodiesel sample, consistent with the method outlined by Demir and Soyhan [14].

2.6 Biodiesel Quality Analysis

To evaluate the effectiveness of the conversion process, acid value determination and biodiesel volume monitoring were conducted at two key stages: after esterification and after transesterification. The acid value, expressed in mg KOH/g, quantifies the amount of potassium hydroxide needed to neutralize the free fatty acids in a given gram of substance, providing a benchmark for assessing fuel quality and its long-term storage stability [15]. It serves as an intermediary for FFA content, which has a direct impact on the transesterification reaction's efficiency and soap formation potential. Successful FFA neutralization is indicated by a significant drop in acid value, which improves the quality of biodiesel and process dependability in general. Standard titrimetric techniques were used to quantify the acid values, and biodiesel quantities were noted at the conclusion of each step to track the development of the yield.

The characterization of biodiesel derived from WVO is essential to ensure compliance with fuel quality standards and environmental regulations. In accordance with the Philippine National Standard for Biofuels – Coconut Methyl Ester B100 (PNS/DOE QS 002:2015), this study focused on three critical parameters: total sulfur content, water and sediment, and FAME (Fatty Acid Methyl Ester) content. These metrics serve as key indicators of biodiesel quality and engine compatibility. Total sulfur content is regulated to mitigate harmful SO_x emissions and extend engine life, while the water and sediment test ensures fuel cleanliness and stability, reducing risks of corrosion and microbial contamination. FAME content, set at a minimum of 96.5% mass by the standard, is vital for verifying the completeness of the transesterification process and ensuring combustion efficiency. All characterization tests in this study were conducted at Intertek Testing Services Philippines, Inc., located in Makati City, Metro Manila, a Department of Energy (DOE)-accredited laboratory for fuel quality analysis in the Philippines.

2.7 Results and Discussion

Figure 1 presents both the acid value reduction and the corresponding biodiesel yield achieved through the microwave-assisted process. The acid value, serving as an indicator of free fatty acid content, was initially measured at 2.11 mg KOH/g in the untreated waste vegetable oil. This was effectively reduced to 0.422 mg KOH/g after acid-catalyzed esterification and further decreased to 0.262 mg KOH/g following transesterification, successfully meeting the Philippine National Standard (PNS) biodiesel specification limit of 0.50 mg KOH/g.

For instance, existing literatuers achieved an acid value reduction from 4.81 to below 0.5 mg KOH/g using microwave-enhanced transesterification in two minutes [16],[17]. Alongside the acid value data, the same figure shows the biodiesel product volumes: from an initial 1.75 L of WVO, the volumes after esterification and transesterification were recorded at 570 mL and 480 mL, respectively, indicating a final biodiesel yield of approximately 27.4%.

Table 1 summarizes the analytical results of the produced biodiesel as evaluated by Intertek Testing Services Philippines, Inc. The sample exhibited a total sulfur content of 0.0007% and a water and sediment content of 0.01%, both of which are well within the acceptable limits set by the Philippine National Standard PNS/DOE QS 002:2015. However, the measured FAME content was 64.8%, which falls below the minimum required value of

96.5%. This lower FAME content corresponds to a biodiesel yield of 64% and may be attributed to the limited separation time of only 9 hours for biodiesel and glycerol layers, a deviation from the conventional 24-hour settling period reported in prior studies [13].

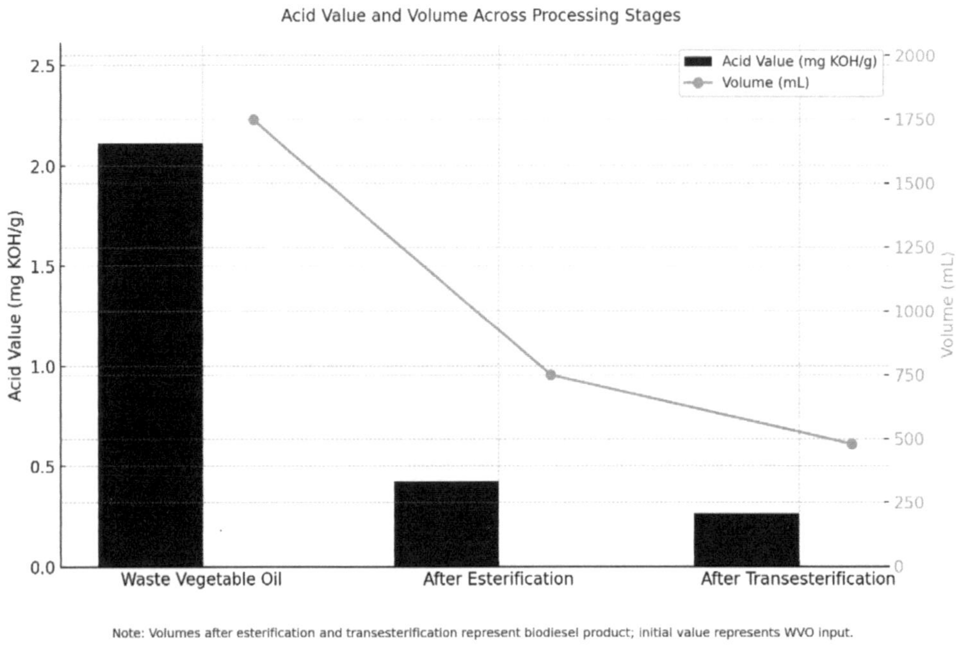

Fig. 1. Acid Value Reduction and Biodiesel Yield at Different Stages of Processing Using Microwave-Assisted Conversion of Waste Vegetable Oil.

Additionally, earlier findings such as those by Babel et al. [8] demonstrated higher conversion efficiencies under more optimized transesterification conditions. These results suggest that further refinement of post-reaction handling and process conditions is necessary to enhance yield and meet the standard specification for biodiesel quality.

Table 1. Biodiesel Characterization Results Based on Philippine National Standards (PNS/DOE QS 002:2015)

TEST CONDUCTED	METHOD	UNIT	SPECIFICATION	RESULT
Total sulfur content	ASTM D2622	% mass	0.001 (max)	0.0007
Water and Sediments	ASTM D2709	% vol	0.05 (max)	0.01
FAME Content	EN14078 mod	% vol	96.5 (min)	64.8

3 Conclusion and Recommendations

This study demonstrated a viable, low cost, and environmentally responsible approach to biodiesel production through the valorization of two locally abundant waste streams: waste vegetable oil and tilapia fish scales. By employing microwave assisted synthesis and reaction technologies, the research achieved substantial reductions in acid value and produced biodiesel with acceptable sulfur and water sediment levels in accordance with PNS/DOE QS 002:2015. Although the measured FAME content did not meet the minimum specification, the observed biodiesel yield and acid value profile validate the technical promise of this dual waste valorization framework. The innovative aspect of the work is the incorporation of waste-derived chitosan as a heterogeneous catalyst, which simplifies the procedure and makes it faster and more sustainable than traditional protocols. These results demonstrate the method's potential scalability for decentralized biofuel projects, especially in areas with limited access to extended reaction infrastructure and conventional catalysts. To further increase FAME content and overall conversion efficiency, process improvement should be given the highest priority moving forward, especially with regard to reaction duration, separation methods, and catalyst reuse. This initiative addresses local environmental management issues while connecting with global sustainability goals and making a significant contribution to the continuing conversation on renewable energy innovation and circular economy practices.

References

1. IQAir, Philippines Air Quality Index (AQI) and Air Pollution information, Aug. 2022
2. S. Sahar, et al., Sustainable Cities Soc. 41, 220 (2018)
3. Zulqarnain, et al., Sustainability 13, 788 (2021)
4. J. Ibáñez, S. Martel Martín, S. Baldino, C. Prandi, and A. Mannu, Processes 8, 798 (2020)
5. F. Musharavati, K. Sajid, I. Anwer, A.-S. Nizami, M.H. Javed, A. Ahmad, and M. Naqvi, Sustainability 15, 16550 (2023)
6. N.A. Hamdi, A. Sulaiman, M.H. Harun, N.F. Mohd Noor, A.S. Ahmad, IOP Conf. Ser.: Earth Environ. Sci. 1091, 012054 (2022)
7. A. Mannu, et al., Waste Manag. 96, 1 (2019)
8. S. Babel, S. Arayawate, E. Faedsura, H. Sudrajat, in Utilization and Management of Bioresources, S. Ghosh (Ed.), Springer, Singapore (2018)
9. A. Kharina, S. Searle, A. D. Rachmadini, A. Kurniawan, A. Prionggo, Int. Council Clean Transport (2018)
10. T.R. Panaligan, J. Aquino, A.M. Alam, J.M. Berroya, P.L.C. Torres, Environ. Sci. Eng. (2025)
11. H. El Knidri, et al., Process Saf. Environ. Prot. 104, 395 (2016)
12. K.V. Supraja, B. Behera, B. Paramasivan, Environ. Sci. Pollut. Res. Int. 22 (2015)
13. U.S. Arachchige, et al., Nat. Environ. Pollut. Technol. 20, 1973 (2021)
14. V.G. Demir, H.S. Soyhan, Eur. J. Eng. Nat. Sci. 2, 137 (2017)
15. P. Onu, C. Mbohwa, in Agricultural Waste Diversity and Sustainability Issues, P. Onu, C. Mbohwa (Eds.), Academic Press, 97 (2021)
16. M.-C. Hsiao, P.-H. Liao, N.V. Lan, S.-S. Hou, Energies 14, 437 (2021)
17. X. Hu, B. Zhang, Q. Wang, Y. Liu, Energies 16, 1741 (2023)

How to reduce environmental burden caused by lignocellulosic waste: a solution from the pharmaceutical industry with green solvents

Lijiao Yu, *Yu* Xia, *Ruiliang* Zhu *and Shun* Yao*

School of Chemical Engineering, Sichuan University, Chengdu, 610065, China

Abstract: In order to reduce environmental burden caused by lignocellulosic waste, lignin from *Eucommia ulmoides* residue was extracted via alkali treatment and acid precipitation, yielding 61.58% purity. Degradation using [N_{4222}]Im and [Ch]Im under atmospheric and microwave conditions was analyzed by elemental, UV, and FT-IR methods. Microwave treatment gave higher hydroxyl (340.54 mg KOH/g) and phenolic hydroxyl (0.32 mmol/g) contents than atmospheric degradation. Gas chromatography analysis showed aromatic compounds accounted for 75.64% of products, dominated by guaiacol and derivatives. This work supports high-value utilization of herbal residues, which is expected to protect our environment.

1 Introduction

With the rapid development of the great health industry, China produces about 70 million tons of herbal residues every year. After extracting the active ingredients, most of the solid residue is naturally degraded by long-term accumulation or directly burned [1]. It has caused significant environmental pollution and waste of resources [2]. Therefore, how to effectively and reasonably utilize them is an unavoidable and important issue in achieving modern production of natural medicines [3]. The energy and harmless utilization of medicinal residues is a necessary way to achieve sustainable development of China's pharmaceutical industry.

The main applications of herbal residues today are composting, edible mushroom cultivation, and feed production [4]. Most composting processes use natural simple composting, but due to the long fermentation cycle and poor hygiene conditions, it is now rarely used; The use of herbal residue for edible mushroom cultivation may inhibit mycelial growth due to the complex composition of the residue, and the separation of the residue is also very difficult. When processing medicinal residues into feed additives, further in-depth and detailed research is needed due to the potential impact of their medicinal functions on animal growth and development. Due to immature technology and limitations, the above utilization methods cannot yet achieve large-scale promotion and application [5].

Similar to some plant resource waste generated in the food industry, herbal residue contains a large amount of components such as crude fiber, crude protein, crude fat, and trace elements [6], which require effective means to achieve their reuse. *Eucommia*

* Corresponding author: cusack@scu.edu.cn

© The Authors, published by EDP Sciences. This is an open access article distributed under the terms of the Creative Commons Attribution License 4.0 (https://creativecommons.org/licenses/by/4.0/).

ulmoides is a precious Chinese medicinal herb commonly cultivated in western, northwestern, and eastern China, belonging to the *Eucommia ulmoides* family. It has the effects of nourishing the liver and kidneys, strengthening muscles and bones, anti hypertension, antioxidant, and anti gastric ulcer [7]. Considering the more and more applications of ionic liquids (ILs) used in the field of biomass in recent years, this study mainly focuses on the degradation of lignin in discarded *Eucommia ulmoides* residue with such green solvents. Firstly, lignin was extracted by traditional acid precipitation and alkali precipitation, and then lignin was degraded by two ILs. The changes in surface hydroxyl groups of lignin before and after degradation and the components degraded into small molecules are measured. At the same time, the difference between extracted lignin and commercialized lignin was also determined. It is expected to provide useful references for reducing waste pollution and resource utilization.

2 Experimental

2.1 Pretreatment

The *Eucommia ulmoides* residue used in this study was obtained after crude extraction of bioactive components from raw material with 95% (v/v) ethanol. The raw residue was soaked in distilled water at a liquid-solid ratio of 20:1 at room temperature to remove residual dust and water-soluble impurities. The suspension was filtered, and the washing procedure was repeated twice. The resulting filter residue was dried in an oven at 50 °C, ground with a laboratory grinder, and sieved through a 40-100 mesh screen. The sieved residue was then mixed with distilled water at a liquid-solid ratio of 50:1, stirred at 60 °C for 1 h, followed by 2 h of standing at room temperature. After filtration, the material was dried in an oven at 50 °C to a constant weight and stored in a desiccator.

2.2 Extraction of Lignin

A 20 g (dry weight) portion of pretreated *Eucommia ulmoides* medicinal residue was mixed with 7% (w/w) sodium hydroxide solution at a liquid-solid ratio of 40:1. The suspension was stirred vigorously at 60 °C for 6 h. Upon completion of the reaction, the mixture was cooled to room temperature and vacuum filtered to remove the cellulose-rich residue. The filtrate was adjusted to pH=2 with 20% (w/w) sulfuric acid and maintained at 50 °C for 1 h to allow lignin precipitation. The suspension was centrifuged at 4000 rpm for 15 min, and the resulting precipitate was collected by filtration, washed repeatedly with distilled water until neutral, and dried at 100 °C to constant weight, yielding a brown solid lignin fraction.

2.3 Purification of Lignin

0.2 g powdered lignin was weighed and placed into a centrifuge tube, followed by the addition of 20 mL of hydrochloric acid solution (pH=3). The suspension was shaken thoroughly to disperse the lignin evenly in the acid solution and soaked for 5 min. The mixture was centrifuged at 4000 rpm, and the supernatant was discarded. This washing procedure was repeated three times with fresh hydrochloric acid solution (pH=3), followed by repeated washing with distilled water until the filtrate reached neutral pH. The residue was dried at 100 °C to constant weight. An accurately weighed portion of the dried product was used for lignin content determination. This treatment effectively removed residual acid-soluble carbohydrates, including cellulose hydrolysis fragments and hemicelluloses trapped or adsorbed in the lignin during precipitation, thereby yielding purified lignin.

2.4 Preparation of ionic liquids (ILs)

According to the screening in pilot experiments, two ILs were successfully synthesized as shown in **Fig. 1**. One was used for the degradation under normal pressure and the other was treated with microwave for comparison.

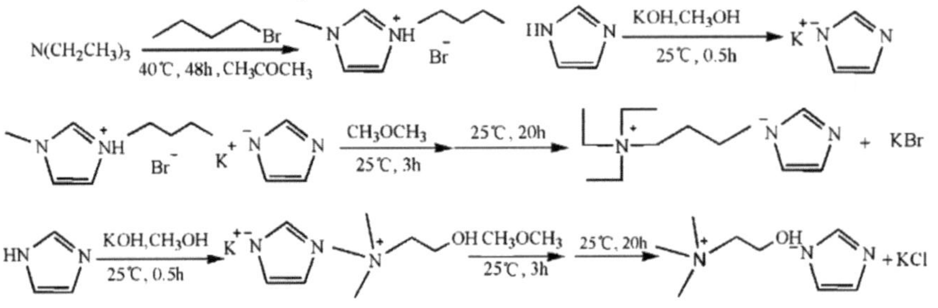

Fig. 1. The synthesis route of ionic liquids [N_{4222}]Im and [Ch]Im.

2.5 Degradation of lignin with [N_{4222}]Im

2.0 g lignin was added to a flask containing 60 mL of distilled water, followed by the addition of 0.1 mol/L ionic liquid and 0.5 mol/L hydrogen peroxide solution. The mixture was reacted at 80 °C for 4 h under normal pressure, after which the reaction was quenched to terminate the reaction. The pH of the solution was adjusted with 1 mol/L hydrochloric acid, and the suspension was centrifuged at 14000 rpm to separate the solid fraction. The recovered solids were vacuum dried at 50 °C to constant weight. The hydroxyl content of the lignin degradation products was subsequently determined.

2.6 Degradation of lignin with [Ch]Im

2.0g lignin was added to a flask containing 60 mL of distilled water, followed by the addition of 0.1 mol/L ionic liquid and 0.5 mol/L hydrogen peroxide solution. The mixture was subjected to 300 W microwave irradiation for 2 h, then rapidly cooled to terminate the reaction. The pH of the resulting solution was adjusted with 1 mol/L hydrochloric acid, and the suspension was centrifuged at 14000 rpm to separate the solid fraction. The recovered solids were vacuum dried at 50 °C to constant weight, and the hydroxyl content of the lignin degradation products was subsequently determined.

2.7 Sample preparation for gas chromatography analysis

Under the optimized conditions, the supernatant obtained by centrifugation at 4000 rpm was extracted three times with 25 mL portions of ethyl acetate. The combined organic layers were dried over anhydrous magnesium sulfate, filtered to remove the drying agent, and concentrated by rotary evaporation at 30 °C. The resulting residue was subsequently analyzed by gas chromatography.

3 Results and discussion

3.1 Analysis of *Eucommia ulmoides* residue

The quantitative method reported in a previous study [8] was employed to determine the contents of cellulose, hemicellulose, and lignin in the pretreated *Eucommia ulmoides*

)ℍt sp:/ odor. :g0ig05g:e3sconf:101229001001

medicinal residue. During cellulose determination using the 72% concentrated sulfuric acid hydrolysis method, a portion of cellulose may undergo hydrolysis and loss, leading to measured values that are approximately 2%–3% lower than the actual content. In lignin measurement, acid hydrolysis of the residue with 72% sulfuric acid effectively removes cellulose and hemicellulose; however, a small fraction of lignin may also be lost, resulting in values that are about 1%–2% lower than the true content. The analytical results are presented in **Fig. 2(a)**.

3.2 Purified lignin content and elemental analysis

The contents of the three components in the purified lignin were also determined using the previously described method, and the results are presented in **Fig. 2(b)**. Following extraction and purification, the lignin content increased significantly from 14.06% to 61.58%. Because both hemicellulose and lignin are alkali-soluble and difficult to separate completely, the relative proportion of hemicellulose remained high, whereas cellulose, being insoluble in both acid and alkali, exhibited the lowest content.

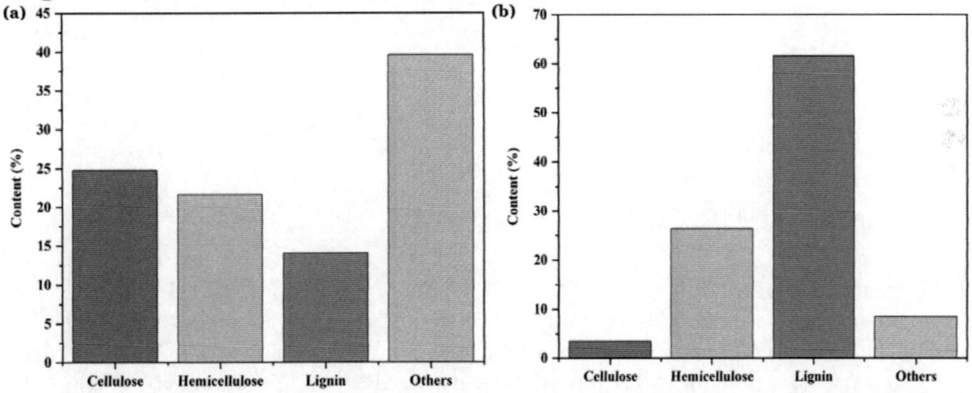

Fig. 2. (a) Composition of *Eucommia ulmoides* medicinal residue after pretreatment; (b) basic composition of purified lignin.

Since elemental analysis cannot directly determine the oxygen content, all elements other than C, H, and N were collectively attributed to oxygen. As shown in **Fig. 3**, the proportions of C, H, and O were relatively consistent. Lignin is primarily composed of C, H, and O, while the nitrogen content is comparatively low, suggesting minimal impurities. The extracted lignin contained 2.15% nitrogen, indicating the presence of a small amount of impurities.

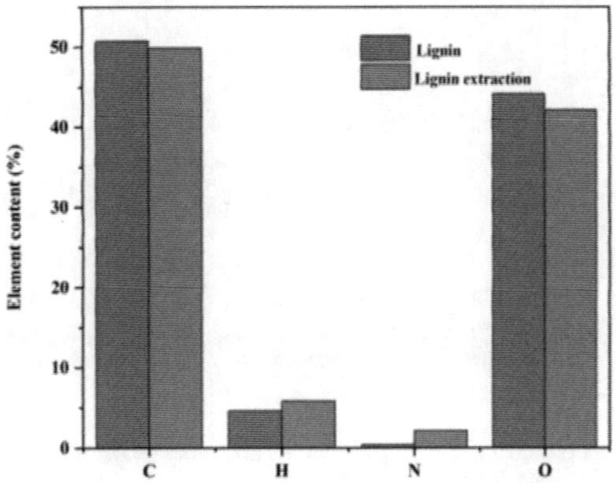

Fig. 3. Elemental composition of purified lignin.

3.3 Spectral analysis of lignin

As shown in **Fig. 4**, the extracted and purified lignin exhibit characteristic absorption peaks at approximately 205 nm and 280 nm. The peak at 280 nm corresponds mainly to the benzene ring and is consistent with the typical spectral features of lignin. However, at the same concentration, the extracted lignin shows a less pronounced peak at 280 nm with lower absorbance. In addition, the absorption band near 205 nm for the extracted lignin exhibits a slight deviation compared with purified lignin, which can be attributed to the relatively high hemicellulose content and incomplete purity of the extracted sample.

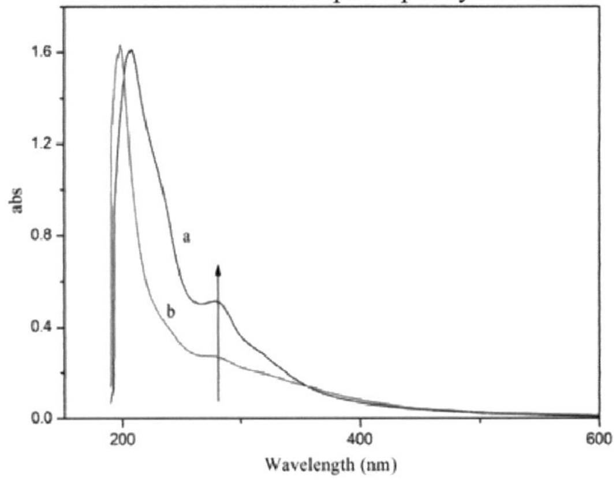

Fig. 4. Ultraviolet (UV) spectra of lignin: (a) native lignin; (b) purified lignin.

As shown in **Fig. 5**, the structural features of lignin are largely consistent with those of the extracted and purified lignin. The O-H stretching vibration of hydroxyl groups appears around 3448 cm^{-1}, while the C-H stretching of aliphatic chains is observed at 3000–2850 cm^{-1}. The C-H stretching of methyl or methylene groups occurs near 2931 cm^{-1}, and the carbonyl (C=O) stretching absorption peak is located at approximately 1718 cm^{-1}. Characteristic absorption bands of the benzene ring are found in the range of 1467–1639 cm^{-1}. In addition, the stretching vibrations of C-C or C-O appear near 1120 cm^{-1}, the C-O stretching of $O\text{-}CH_3$ or C-OH groups is observed around 1047 cm^{-1}, and the aromatic C-H bending vibration occurs at approximately 900 cm^{-1}. Compared with lignin, the purified lignin exhibits weaker benzene ring absorption within 1467–1639 cm^{-1}, indicating a lower aromatic ring content [9]. The ether bond content is also relatively reduced.

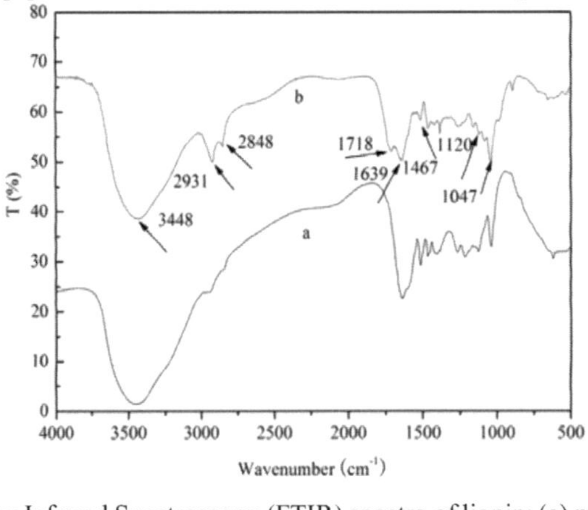

Fig. 5. Fourier-Transform Infrared Spectroscopy (FTIR) spectra of lignin: (a) native lignin; (b) purified lignin.

3.4 Characterizations on lignin degradation products

Based on the ionic liquid degradation method described earlier, elemental analysis was performed to examine the changes in C, H, O, and N before and after degradation, thereby further verifying the structural transformations of lignin. As shown in **Fig. 6**, the C and N contents of lignin increased after degradation, while the O content decreased. The increase in C content may be attributed to the stability of aromatic rings, as the C–C bonds are less susceptible to cleavage compared with ether bonds. Meanwhile, other functional groups or side chains may be degraded into smaller molecules, leading to a higher degree of C–C condensation and consequently elevated C content. The decrease in O content can be explained by the cleavage of methoxy groups during the reaction. The higher N content likely results from the presence of non-degradable N-containing impurities, which increase the relative proportion of N after the reaction. Furthermore, elemental analysis of atmospheric-pressure and microwave degradation products revealed a significant increase in H content, indicating that ether bond cleavage generated additional hydroxyl groups. This also explains the higher hydroxyl content observed in lignin degraded by microwave compared with atmospheric-pressure degradation. The reduction in O content further suggests that microwave-assisted degradation proceeds to a greater extent.

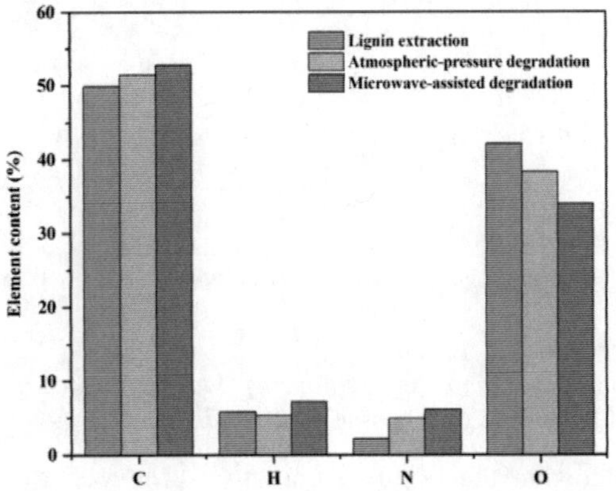

Fig. 6. Elemental composition of degradation of purified lignin and raw lignin.

As shown in **Fig. 7**, after lignin degradation, the absorption peak at 205 nm remains largely unchanged, whereas the peak at 280 nm is no longer pronounced, and the overall absorbance in the higher-wavelength region decreases significantly. These changes indicate that the conjugated structure of lignin has been substantially disrupted.

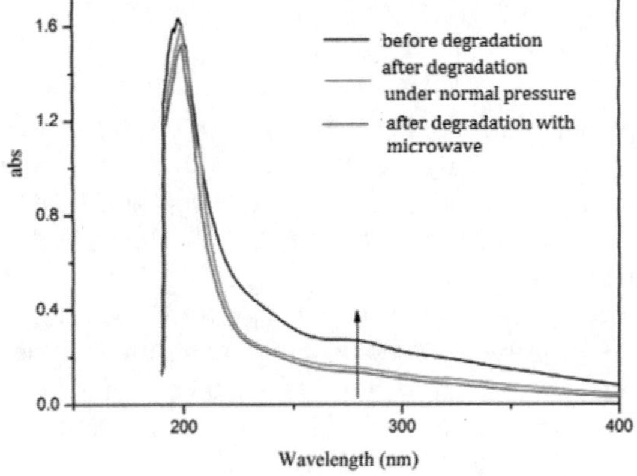

Fig. 7. UV spectra of degraded lignin.

As shown in **Fig. 8**, the FTIR spectra of degraded lignin are generally consistent with those of lignin prior to degradation. The O-H stretching vibration of hydroxyl groups appears around 3437 cm^{-1}, C-H stretching of methyl or methylene groups near 2925 cm^{-1}, C-H stretching of methoxy groups around 2850 cm^{-1}, and the carbonyl (C=O) stretching vibration at approximately 1709 cm^{-1}. Characteristic absorption bands of the benzene ring are observed between 1467–1639 cm^{-1}, C-C or C-O stretching near 1126 cm^{-1}, C-O stretching in O-CH_3 or C-OH groups around 1045 cm^{-1}, and aromatic C-H bending at approximately 900 cm^{-1}. Notably, the microwave-degraded lignin exhibits the highest intensity at 1126 cm^{-1}, corresponding to the in-plane C-O vibration of guaiacol. This increased intensity indicates the formation of a substantial guaiacol structure during degradation, which also explains the higher surface hydroxyl content observed after microwave-assisted treatment.

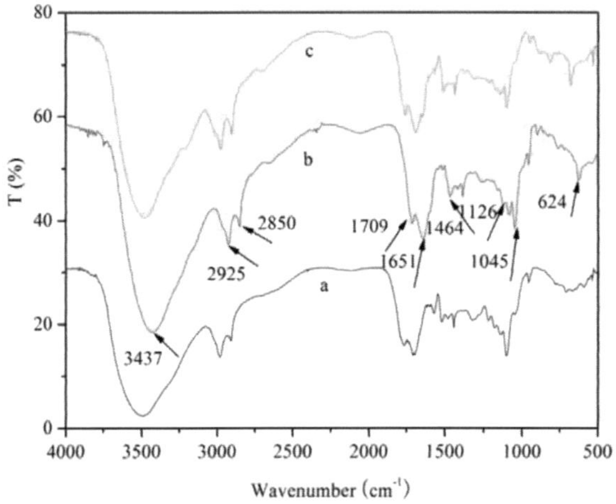

Fig. 8. FTIR spectra of lignin: (a) before degradation, (b) after degradation under atmospheric pressure, (c) after microwave-assisted degradation (KBr pellet).

GC analysis revealed that the lignin degradation products primarily consist of acids, alcohols, aldehydes, esters, and other small molecules. Aromatic compounds account for approximately 75.64% of the total products, with guaiacol representing about 30.89% of the aromatics and 2-methoxy-4-vinylphenol reaching 38.38%, indicating that guaiacol-based monomers are the predominant species in *Eucommia ulmoides* residue. Structural analysis further confirms that the degradation mainly targets the cleavage of β-O-4 ether bonds [10].

4 Conclusions

Lignin was extracted and purified from *Eucommia ulmoides* medicinal residue under optimized conditions, and the hydroxyl content of the resulting oligomeric lignin, as well as the composition of derived small-molecule products, was analyzed. The residue contained 24.77% cellulose, 21.62% hemicellulose, and 14.06% lignin, while the purified lignin reached 61.58% content.

Characterization by elemental analysis, UV spectroscopy, FTIR, and other techniques showed that the structural features and elemental composition of the extracted lignin were largely consistent with commercial lignin. Degradation under atmospheric-pressure and microwave-assisted conditions indicated that microwave treatment was more effective, producing higher aliphatic (340.54 mg KOH/g) and phenolic (0.32 mmol/g) hydroxyl contents. GC–MS analysis revealed that aromatic compounds, particularly guaiacol-based monomers, predominated among the degradation products, confirming efficient cleavage of lignin structures.

From an industrial perspective, the abundant and low-cost *Eucommia ulmoides* residue represents a sustainable feedstock for lignin valorization. The combination of established extraction and purification processes with microwave-assisted degradation enables scalable production of small-molecule aromatic compounds, which can serve as precursors for bio-based chemicals, resins, and functional polymers. This strategy provides a green, efficient, and practical approach for high-value industrial applications of lignin.

Special thanks to the support from National Natural Science Foundation of China (82273814) and Fundamental Research Funds for the Central Universities/Sichuan University-Luzhou Science and Technology Innovation Platform Construction Project (2022CDLZ-20). We would like to thank Xiang Lin from Center of Engineering Experimental Teaching, School of Chemical Engineering, Sichuan University for FTIR.

References

1. X. Long, Y.-L. Lu, H. Guo, Y.-P. Tang, ChemistrySelect, **8**, e202300383, (2023).
2. C. Huang, Z.-X. Li, Y. Wu, Z.-Y. Huang, Y. Hu, J. Gao, J. Environ. Manage. **299**, 113607 (2021).
3. X. Wu, J. Wang, C. Amanze, R. Yu, J. Li, X. Wu, L. Shen, Y. Liu, Z. Yu, W. Zeng, J. Environ. Manage. **319,** 115765 (2022).
4. W. Tao, J. Jin, Y. Zheng, S. Li, Waste Biomass Valori. **12,** 5853-5868 (2021).
5. B. Wu, T. Ren, X. Cao, T. Wu, Z. Hu, J. Ai, N. Zhang, Y. Zhang, Z. Yu, L. Du, H. Niu, Ind. Crops Prod. **227** 120802 (2025).
6. Q. Lu, C. Li, J. Cleaner Prod. **279,** 123856 (2021).
7. T. Hussain, B. Tan, G. Liu, O. A. Oladele, N. Rahu, M. C. Tossou, Y. Yin, Evid. Based Complement. Alternat. Med. **2,** 5202908 (2016).
8. S. Yao, Y. Yang, H. Song, Y. Wang, H. Wan, J. Sci. Ind. Res. **74**, 58-63 (2015).
9. S. Zhang, J. Xiao, G. Wang, G. Chen, Bioresource Technol. **304,** 122975 (2020).
10. Y. Zhang, F. Huo, Y. Wang, Y. Xia, X. Tan, S. Zhang, H. He, Front. Chem. **7,** 78 (2019).

Fabrication of Eco-Friendly Pineapple Leaf Fiber-Based Vegan Leather for Environmental Sustainability

Siriwan Srisang, *Jumpon* Kunyuan, *Thunyanat* Hutangkoon, *Orranat* Eiangmee, *Jatuphat* Maikaew, *Siwakon* Sokjabok* and *Naruebodee* Srisang

Department of Engineering, King Mongkut's Institute of Technology Ladkrabang, Prince of Chumphon Campus, Chumphon 86160, Thailand

Abstract. The environmental impact of synthetic leather production has raised global concerns due to its reliance on petroleum-based polymers and poor biodegradability. Therefore, the development of sustainable, eco-friendly alternatives using renewable resources has become increasingly important. A biodegradable vegan leather was developed from natural rubber and pineapple leaf fibers (PALF), with properties analyzed using Response Surface Methodology (RSM). The effects of fiber content (X1), compression time (X2), and compression temperature (X3) were studied on biodegradation (Y1), water absorption (Y2), and tensile strength (Y3). Results showed that all three factors significantly influenced Y1, with the predictive model demonstrating high reliability (R > 90%). The optimum condition for Y1 was X1 = 3.0 g, X2 = 70.0 min, and X3 = 110.0 °C, yielding a maximum predicted biodegradation of about 21%. In contrast, the models for Y2 and Y3 were statistically unreliable ($P > 0.05$) due to low R^2 values. However, Y2 passed the lack-of-fit test, suggesting an adequate model form, while Y3 failed ($P < 0.05$), indicating an inadequate prediction model. These findings suggest future experiments should narrow factor ranges and include additional control variables to improve the predictability of Y2 and Y3. Despite these limitations, the study highlights a sustainable alternative to conventional synthetic leather, aligning with circular economy principles and supporting the United Nations Sustainable Development Goals (SDGs). Importantly, the process is resource-efficient: from 1 kg of pineapple leaves, only 20 g of fibers is obtained, and just 75 g of PALF was used in 15 experimental runs. This minimal material requirement underscores the potential of the approach for sustainable production, while highlighting potential applications in sustainable fashion, packaging, and eco-friendly product design.
Keywords: Vegan leather, Natural rubber, Pineapple leaf fiber, Response surface methodology, Circle Economy and SDGs

1 INTRODUCTION

The 2030 Agenda for Sustainable Development, announced by the United Nations in 2015, established 17 Sustainable Development Goals (SDGs) to address critical global challenges, including poverty, inequality, injustice, and climate change.

*Corresponding author: siwakon.so@kmitl.ac.th

© The Authors, published by EDP Sciences. This is an open access article distributed under the terms of the Creative Commons Attribution License 4.0 (https://creativecommons.org/licenses/by/4.0/).

Among these, SDG 12 emphasizes responsible consumption and production, promoting a transition from a linear economic model to a circular economy. The valorization of agricultural wastes and by- products presents significant opportunities for sustainable development by converting residues into valuable products, offering both ecological and socio-economic benefits. In this context, the bioeconomy refers to the recycling and valorization of renewable agro-industrial waste into novel products, such as animal feed, food, chemicals, energy, and bio-based materials, whereas the circular economy aims to close the loop of production and consumption [1]. In the food industry, large quantities of by-products can serve as renewable resources when processed using green technologies. For instance, fish farming a rapidly growing sector, utilizes only 30-40% of the fish as food, leaving 60-70% as by-products. One valorization approach is upcycling fish skin into leather [2]. The increasing industrial demand for animal leather in the fashion industry has driven research into alternative materials, such as vegan or artificial leather, produced from natural or synthetic sources. These materials mimic the appearance of animal leather without using animal skins. While the leather industry has expanded over recent decades, the associated environmental impacts have also grown due to greenhouse gas (GHG) emissions primarily originating from animal-based feedstock. Developing bio-based materials, containing at least one biologically produced and fully biodegradable component, is a promising strategy to reduce environmental impacts [3]. Ethical awareness among consumers has led to growing interest in vegan materials, driven by concerns for animal welfare, environmental sustainability, and ethical production practices [4]

Agriculture plays a significant role globally, with approximately 25% of senior executives involved in agriculture and land management. An example, as illustrated in the Fig.1, can be seen in the pineapple cultivation areas of Prachuap Khiri Khan Province, Thailand. Thailand, as one of the world largest agricultural producers, is a key hub for tropical crop production. Pineapple is widely cultivated in tropical and subtropical regions, including Thailand, Malaysia, and the Philippines. According to the Food and Agriculture Organization (FAO), global pineapple production reached over 28.18 million tons in 2018 and has continued to grow over the past decade. Processed pineapple products such as canned pineapple, juice, and juice concentrate constitute major exports, with Thailand ranked as the largest exporter of canned pineapple in 2018. Recent FAO data indicate that global pineapple exports are projected to grow by approximately 4% in 2023 [5]. Despite economic benefits, intensive pineapple cultivation generates large volumes of agricultural waste, including leaves, crowns, peels, and cores. Much of this waste is discarded or openly burned, contributing to greenhouse gas emissions and air pollution, particularly PM2.5, which adversely affects environmental quality and public health. Therefore, responsible agricultural waste management and the development of value-added products from pineapple residues are urgently needed. Utilizing pineapple leaf fibers (PALF) and other by-products for innovative applications such as textiles, bio-composites, and vegan leather can reduce environmental impacts, lower disposal costs, generate additional income for farmers, and support the transition toward a sustainable circular economy [6-7].

However, previous studies on bio-based leather alternatives have primarily focused on either synthetic polymers or single-source natural fibers, often overlooking the synergistic potential of combining natural rubber with agricultural residues such as PALF. Moreover, few studies have systematically optimized processing parameters to simultaneously maximize mechanical performance, water resistance, and biodegradability using statistical design tools like RSM. This study addresses these gaps by developing and optimizing a PALF natural rubber composite leather substitute, aiming to achieve a balance between sustainability, functionality, and scalability for potential industrial applications.

Fig. 1. Pineapple leaves from Prachuap Khiri Khan Province

2 LITERATURE REVIEW

With the world warming, habitats destroyed and synthetic materials impacting the environment, effective waste sorting is essential, with monitoring stations in Surat Thani and Songkhla provinces, key southern hubs in Thailand, being assessed for their overall waste volume before it is released into the ocean. Waste classification is therefore crucial [8]. The production of vegan leather, an invention for sustainable fashion policy, is created by using waste materials instead of animal skins and furs to produce expensive fashion clothing. The leather that will replace animal skins will be artificial leather from the polyurethane and polyester groups and leather plants-based; each group has different advantages and disadvantages [9]. In artificial leather production, mechanical properties play a crucial role in determining suitable materials for various applications and influencing consumer preferences. Consequently, classifying artificial leather types based on their mechanical properties has been widely studied, as each type exhibits unique characteristics that impact user preferences and market sustainability. Therefore, the production process must carefully consider these characteristics and material compositions to ensure both functionality and long-term demand. [10]. There is also a study that investigated the effects of outdoor exposure on the tensile properties and water absorption behavior of synthetic leather made from polyurethane (PU) and polyester. The experiments were conducted under general environmental conditions at room temperature, following the Japanese Industrial Standard (JIS) Z 2381. In addition, the study considered optical degradation under outdoor exposure. The results indicated that the tensile properties of synthetic leather are primarily influenced by optical degradation caused by ultraviolet (UV) irradiation during outdoor exposure. [11]. The foundation for developing eco-friendly artificial leather focuses on creating sustainable alternatives in the synthetic leather industry through the recycling of organic waste materials. Previous studies have explored the utilization of fruit peels, including pomelo, Sunkist orange, and grapefruit, as components in artificial leather production. These studies primarily emphasized the mechanical properties of the resulting material, reporting a tensile strength of 472 N/cm², an elongation value of 67.28%, and a tear resistance of 14.28 N/cm². These findings highlight the potential of plant-based materials as viable and sustainable alternatives to animal-derived leather [12]. There has been a study on the treatment of PALF with liquid smoke and its effects on both tensile strength and impact strength. The results indicate that liquid smoke treatment can enhance the strength and toughness of PALF. The tensile strength test was conducted in accordance with the ASTM D638 Type IV standard. The study showed that a 1-hour immersion resulted in a tensile strength of 64.42 MPa, which increased to 72.53 MPa after 2 hours, and further increased to 74.65 MPa after 3 hours of immersion. For the impact strength test, which followed the ASTM D5942 standard, the composite reinforced with PALF exhibited an impact strength of 9.32 J/m² after 1 hour of treatment. This value increased to 13.44 J/m² after 2 hours but decreased to 10.21 J/m² after 3 hours of immersion.

These findings demonstrate that liquid smoke treatment can significantly improve the mechanical properties of PALF, with the most notable enhancement in tensile strength occurring at 3 hours and the highest impact strength observed at 2 hours of treatment [13]. There has been a study on the development of plant-based leather alternatives and natural rubber (NR)-based leather substitutes reinforced with PALF. A simple process was employed, in which PALF was extracted from pineapple leaf waste using a mechanical method. Both untreated PALF (UPALF) and PALF treated with sodium hydroxide (TPALF) were fabricated into fiber mats using a paper-making process. Subsequently, the PALF mats were coated with natural rubber latex at three different NR/PALF ratios: 60/40, 50/50, and 40/60. The resulting bio composite leathers were then evaluated for their tensile strength, tear resistance, and hardness. In addition, the internal microstructure of the leather was observed using a scanning electron microscope (SEM) and compared across samples. The results showed that the leather with an NR/PALF ratio of 50/50 produced the most satisfactory outcome. Therefore, this type of alternative leather demonstrates unique characteristics, being a biobased leather with a lower carbon footprint [14].

To raise awareness within the community, promote the utilization of agricultural waste from local harvests, and support local waste management, the aim is to maximize the value of all materials and ensure their sustainable use. This also encourages the sustainable use of natural rubber and creates opportunities for future development. Furthermore, previous studies have not considered statistical analysis using response surface methodology (RSM). In this work, RSM is applied through a Box–Behnken experimental design to enhance environmental sustainability

3 MATERIALS AND METHODS

This research aimed to investigate the fabrication process of bio-based artificial leather using natural rubber latex and PALF. The study applied RSM through a Box–Behnken design to optimize three critical properties: biodegradability (Y_1), water absorption (Y_2), and tensile strength (Y_3). Three key factors were considered: the amount of PALF (X_1: 3 g, 5 g, 7 g), The compression temperature (X_2: 70 °C, 90 °C, 110 °C) was monitored using a FLIR E60 non-contact infrared thermometer, which is specifically designed for high-temperature measurement without direct sample contact and provides a thermal resolution of 320 × 240 pixels, and compression time (X_3: 10 min, 20 min, 30 min) as illustrated in Fig 2.

Besides, these parameters were selected based on their significant influence on the structural integrity and functional performance of the final composite material. The variables were set at three levels: low (-1), medium (0), and high (+1), in order to comprehensively analyse and determine the effects of each factor on the material properties. By employing the Box–Behnken Design (BBD) in conjunction with statistical response analysis, the main objective was to identify the combination of the three variables that would result in a composite material with the desired properties in terms of biodegradability, water absorption, and tensile strength. These properties are crucial to developing sustainable and environmentally friendly alternative materials, as summarized in Table 1. and the experimental design generated 15 experimental runs as summarized in Table 2, enabling statistical modeling and optimization to identify the ideal combination of processing conditions that balances mechanical performance with environmental sustainability.

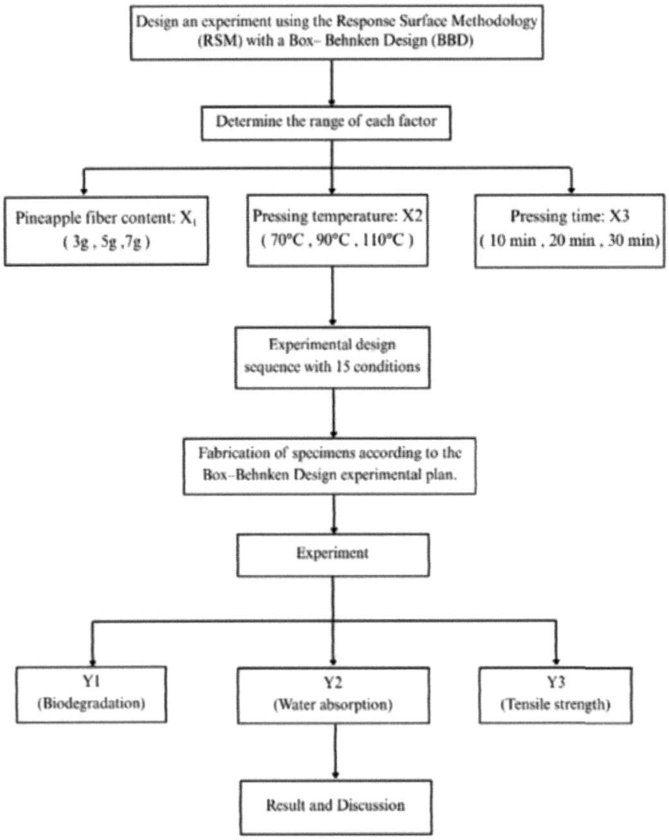

Fig. 2. Flow chart presents studying that sustain for Pineapple Leaf Fiber-Based Vegan Leather

Fig. 3. The main component process began of bio-based artificial leather using natural rubber latex and pineapple leaf fibers.

Table 1. Design Variable Levels

Type	Code	level		
		-1	**0**	**1**
Amount of pineapple leaf fibers	X_1	3	5	7
Compression temperature	X_2	70	90	110
Compression time	X_3	10	20	30

Table 2. Experimental sequence from the design of Box-Behnken

Numbers	X_1	X_2	X_3
1	3	70	20
2	7	70	20
3	3	110	20
4	7	110	20
5	3	90	10
6	7	90	10
7	3	90	30
8	7	90	30
9	5	70	10
10	5	110	10
11	5	70	30
12	5	110	30
13	5	90	20
14	5	90	20
15	5	90	20

As illustrated in Fig.3, the process began with the preparation of pineapple leaves through chemical treatment to obtain semi-dry fibers suitable for forming. These fibers were fabricated into paper-like sheets measuring 28 × 28 cm at weights of 3 g, 5 g, and 7 g, with the weights precisely measured using a digital scale (PR224/E, OHAUS), according to the experimental design. Subsequently, pre-vulcanized natural rubber latex (MM Prevulcanized latex) was prepared at a volume of 70 mL for each run. The latex was then poured into aluminum trays of the same dimensions (28 × 28 cm) and processed using a MEMMERT UF 30 oven under specified experimental conditions. After completing the forming process according to the experimental design table, the samples were tested to evaluate the three properties: Y_1, Y_2, and Y_3.

In this study, experiments were conducted to determine the optimal conditions for fabricating bio-based composite materials by employing an experimental design approach, which is one of the techniques of Response Surface Methodology (RSM). This method is widely recognized in engineering and materials science research because it effectively reduces the number of experimental runs without compromising the accuracy of the results. Moreover, it allows for a clear analysis of the interactions between variables. In this experimental design, three independent variables were considered. And the molding process was carried out along with temperature monitoring. The molding was performed using a hydraulic press (TOYO 30 tron) with precise timing control, as shown in Fig.4, while the preparation procedure of the artificial leather is illustrated in Fig. 5.

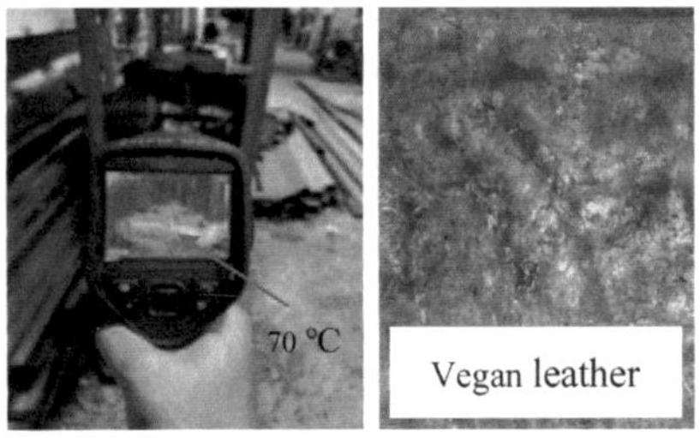

Fig. 4. The molding was performed using a hydraulic press under controlled temperature conditions.

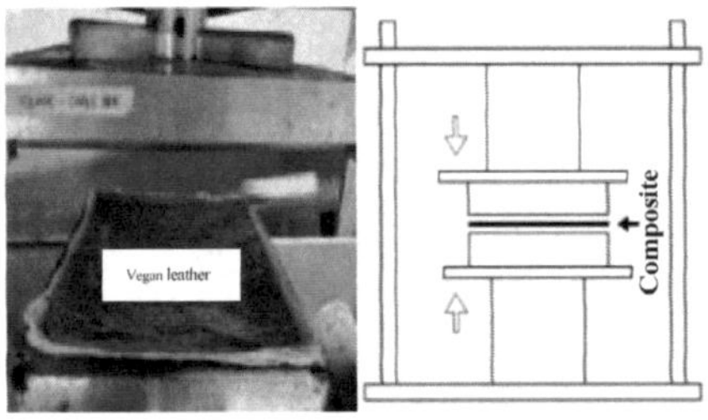

Fig. 5. Preparation procedure of the vegan leather

A. Biodegradation

The results were analyzed following the ASTM G160 standard. Specimens were cut into dimensions of 2 cm × 2 cm and dried at 60 °C for 24 hours [15]. Each specimen was weighed prior to testing, then buried in soil at a depth of 2 cm [16] for a duration of 14 days. After the burial period, the samples were retrieved, cleaned, dried again, and reweighed to determine the weight loss. A comparison of the specimens before and after 14 days of biodegradation testing is shown in Fig. 6.

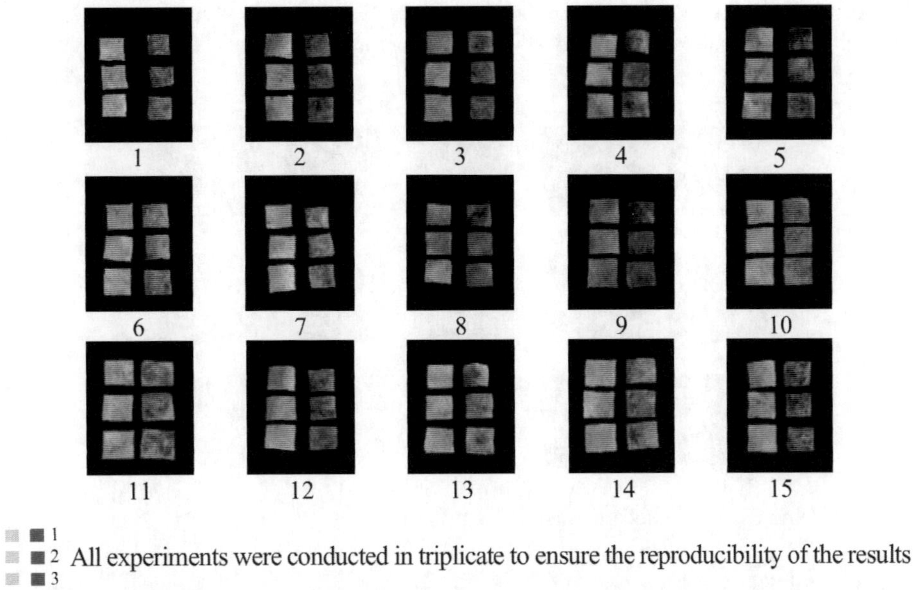

Fig. 6. Comparison of samples before and after biodegradation testing at 14 days. (The experiments were repeated three times to verify the consistency of the results)

B. Water absorption

The experimental results were analyzed using specimens measuring 2 cm × 2 cm in accordance with ASTM D570. Each specimen was weighed prior to testing, then dried at 60 °C for 24 hours. The samples were subsequently immersed in 10 mL of DI water for 24 hours. Excess water was removed using absorbent paper, and the specimens were weighed again [17], as illustrated in Fig. 7.

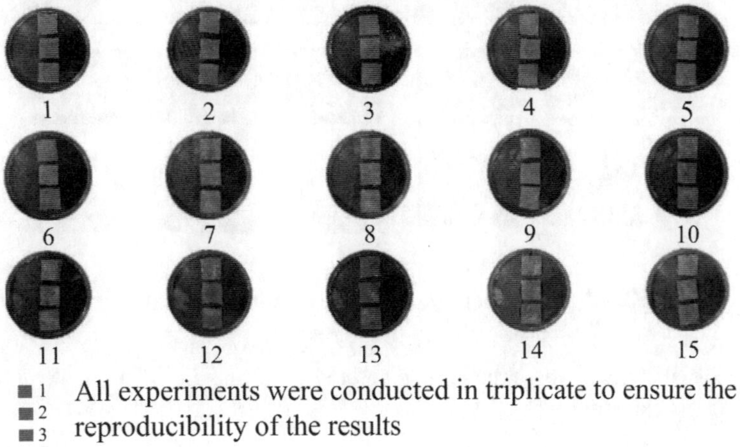

Fig. 7. The samples were tested for water absorption over a 24 hr. period

) Ht sp:/ odor. :g0ig05g:e3sconf:101229001003

C. Tensile strength

The test specimens were analyzed according to ASTMD412 and ISO 37 standards, with the samples cut into a dumbbell shape (Type C). Tensile testing was performed at a constant crosshead speed of 100 mm/min with a load of 1 kg [18]. The tests were conducted at room temperature using a Universal Testing Machine, as illustrated in Fig. 8.

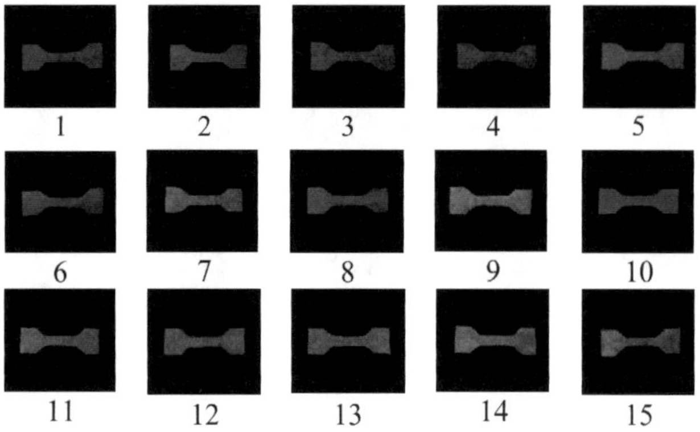

Fig. 8. Type C sample before tensile strength testing.

D. UV aging

The analysis focused on the degradation of materials when exposed to ultraviolet (UV) radiation from sunlight over extended periods. This is one of the critical factors contributing to the deterioration of various materials, particularly plastics, rubber, and polymers, which may experience a loss of properties. The analysis involved observing color fading, discoloration, or changes in surface appearance, such as embrittlement or surface cracking. The exposure tests were conducted for durations of 24 hours, 48 hours, and 72 hours [19], as illustrated in Fig. 9.

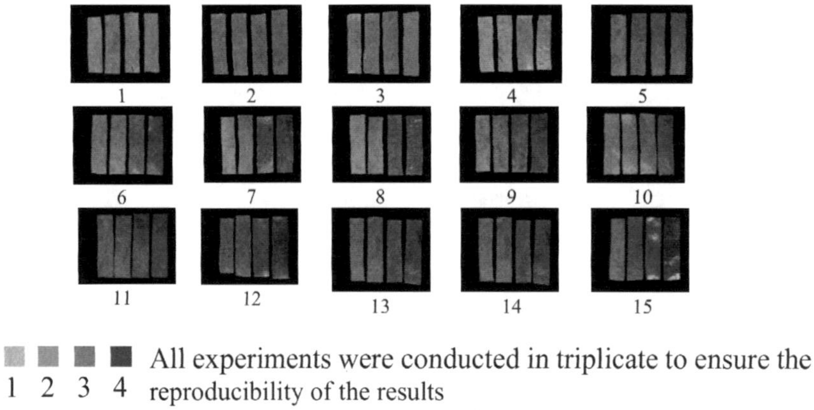

Fig. 9. Visual comparison of the sample before UV testing and after 24, 48, and 72 hours of UV exposure.

4 RESULT AND DISCUSSION

This study, the objective was to determine the optimal conditions for fabricating artificial leather combined with natural rubber. The experimental design was based on the Response Surface Methodology (RSM) using Minitab 21 software, investigating three factors at three levels: Y_1, Y_2, and Y_3. The fabricated samples were then subjected to property testing to evaluate their performance and the efficiency results are presented in Table 3.

The fabricated specimens were subjected to comprehensive property evaluations to assess their overall performance. The experimental results were subsequently analyzed using statistical modeling to derive regression equations describing the influence of each variable on the responses. The analysis encompassed the estimation of regression coefficients to quantify factor effects, lack-of-fit testing to verify model adequacy, determination of coefficients of determination (R^2) and adjusted R^2 to evaluate model fit, and predictive indicators to assess the model's forecasting capability. Finally, numerical optimization was performed to identify the most suitable combination of variables, achieving optimal material properties while minimizing experimental cost. The results of the experiment were divided as follows

Table 3. The efficiency results

	Factors			Response Variables		
No.	X_1(g)	X_2(°C)	X_3(min)	Y_1(%)	Y_2(%)	Y_3(MPa)
1	3	70	20	19.70	15.79	1.12
2	3	110	20	11.41	16.97	1.58
3	7	70	20	11.36	18.23	2.35
4	7	110	20	7.86	12.50	3.44
5	3	90	10	17.41	17.48	1.48
6	3	90	30	4.21	19.64	1.82
7	7	90	10	12.15	21.82	2.93
8	7	90	30	9.05	14.68	3.46
9	5	70	10	14.51	13.33	1.45
10	5	70	30	10.20	11.37	1.86
11	5	110	10	19.35	12.85	2.79
12	5	110	30	13.12	12.84	3.11
13	5	90	20	14.86	10.90	2.44
14	5	90	20	12.88	12.96	2.46
15	5	90	20	10.88	16.94	2.51

A. Analysis of Variance - ANOVA of biodegradation.

From the study, the biodegradation test results were collected after 14 days, as shown in Fig.6, comparing the samples before and after the test, with three replications. In addition, SEM images were taken before and after the test at magnifications of 400X and 20X to observe the surface morphology of the artificial leather and to evaluate the biodegradation efficiency after soil burial, Subsequently, a response surface regression analysis was performed to determine the relationship among the factors, based on statistical probability values (P-values) at a 95% confidence level. If the P-value was less than or equal to 0.05, the variable was considered statistically significant in the model for biodegradation testing. The results indicated that the significant factors were the X_1 with a P-value of 0.004 and X_2 with a P-value of 0.010, while X_3 had a P-value of 0.241, indicating no significant effect.

Furthermore, the lack-of-fit value was found to be 0.553, indicating that the model adequately fits the data. When considering the coefficient of determination (R^2) of 92.24% and the adjusted of 78.27%, it suggests that the model could explain a substantial proportion of the variability in the response and the ANOVA results are presented in Table 4.

After considering the P-value, a multiple regression equation was created, as shown in Eq.1

$$\text{Biodegradation (\%)} = 74.6 - 2.12 X_1 - 0.984 X_2 - 0.232 X_3 - 0.485X_1^2 + 0.00412X_2^2 - 0.0023X_3^2 + 0.2991X_2 + 0.1263 X_1 X_3 - 0.00240X_2X_3$$

B. *Analysis of Variance - ANOVA of water absorption.*

Water absorption reflects the ability of artificial leather to absorb moisture or liquid from the surrounding environment, particularly in plant-fiber-based materials with a porous structure. The results were analyzed from water absorption tests conducted over a 24-hour period to evaluate the material's performance in maintaining its integrity when exposed to water. Type sizes for final papers. The analysis focused on the variation of the model and examined the relationship between factors. Statistical significance was determined using the P-value, where a P-value less than or equal to 0.05 indicates that the variable is accepted in the model and has a statistically significant effect on water absorption.

From the analysis of variance (ANOVA), it was found that X_1 had a P-value of 0.146, X_2 had a P-value of 0.503, and X_3 had a P-value of 0.949. The lack-of-fit value was 0.998, indicating that the experimental data were highly suitable for the model. Furthermore, the quadratic term of X_1^2 had a P-value of 0.011, showing statistical significance. The coefficient of determination (R^2) was 86.73%, which indicates that the model explains a high proportion of the variation in the data. From the results, the independent variable explained 13.27% of the variation, indicating limited control. The R-sq (adj) was 62.85% suggesting that the model could explain a substantial portion of the variance in the data. However, the R-sq (pred) was 0.00%, showing that the model had no predictive ability for water absorption. This may be due to the unclear relationship with tensile strength or an insufficient amount of data and the ANOVA results are presented in Table 4.

C. *Analysis of Variance - ANOVA of tensile strength*

The tensile test is a method used to evaluate the ability of a material to resist forces that attempt to stretch it. The results are recorded for subsequent analysis of the material's strength.

From the ANOVA, it was found that X_1 had a P-value of 0.084, compression X_2 had a P-value of 0.019, and X_3 had a P-value of 0.004. The lack-of-fit value was 0.005, indicating that the experimental data were not suitable for the model. The coefficient of determination (R^2) was 89.84%. While, the R-sq (adj) was 71.55% suggesting that the model could explain a substantial portion of the variance in the data. However, the R-sq (pred) was 0.00%, showing that the model had no predictive ability for water absorption. This may be due to the unclear relationship with Tensile strength or an insufficient amount of data.

For the results to be reliable, the experiments should be repeated to obtain more consistent values, and other relevant factors should be considered to accuracy and the ANOVA results are presented in Table 4.

Table 4. ANOVA for response surface methodology on the efficiency results of pineapple leaf fiber-based vegan Leather

Source	Y_1[a]					Y_2[b]					Y_3[c]				
	DF	Adj SS	Adj MS	F-Value	P- Value	DF	Adj SS	Adj MS	F-Value	P- Value	DF	Adj SS	Adj MS	F-Value	P- Value
Model	9	226.495	25.1661	6.60	0.026	9	125.543	13.9492	3.63	0.085	9	6.96595	0.77399	4.91	0.047
Linear	3	168.254	56.0845	14.72	0.006	3	13.370	4.4566	1.16	0.411	3	6.58890	2.19630	13.94	0.007
X_1	1	98.631	98.6310	25.88	0.004	1	11.353	11.3526	2.96	0.146	1	0.73205	0.73205	4.65	0.084
X_2	1	62.888	62.8881	16.50	0.010	1	2.000	2.0000	0.52	0.503	1	1.82405	1.82405	11.58	0.019
X_3	1	6.734	6.7345	1.77	0.241	1	0.017	0.0171	0.00	0.949	1	4.03280	4.03280	25.59	0.004
Square	3	26.081	8.6938	2.28	0.197	3	77.663	25.8877	6.74	0.033	3	0.26677	0.08892	0.56	0.662
X_1^2	1	13.902	13.9023	3.65	0.114	1	60.264	60.2644	15.69	0.011	1	0.04777	0.04777	0.30	0.606
X_2^2	1	10.047	10.0472	2.64	0.165	1	11.535	11.5350	3.00	0.144	1	0.20174	0.20174	1.28	0.309
X_3^2	1	0.192	0.1918	0.05	0.831	1	2.161	2.1608	0.56	0.487	1	0.01621	0.01621	0.10	0.761
2-Way Interaction	3	32.160	10.7200	2.81	0.147	3	34.510	11.5034	3.00	0.134	3	0.11028	0.03676	0.23	0.870
$X_1 X_2$	1	5.736	5.7360	1.51	0.274	1	11.937	11.9370	3.11	0.138	1	0.09923	0.09923	0.63	0.463
$X_1 X_3$	1	25.502	25.5025	6.69	0.049	1	21.622	21.6225	5.63	0.064	1	0.00902	0.00902	0.06	0.820
$X_2 X_3$	1	0.922	0.9216	0.24	0.644	1	0.951	0.9506	0.25	0.640	1	0.00203	0.00203	0.01	0.914
Error	5	19.055	3.8109			5	19.203	3.8406			5	0.78785	0.15757		
Lack-of Fit	3	11.134	3.7115	0.94	0.553	3	0.348	0.1159	0.01	0.998	3	0.78525	0.26175	201.35	0.005
Pure Error	2	7.920	3.9601			2	18.855	9.4276			2	0.00260	0.00130		
Total	14	245.550				14	144.746				14	7.75380			
S = 1.95216						S = 1.95973					S = 0.396951				

[a]Y_1: Biodegradation , [b]Y_2: Water absorption ,[c]Y_3: tensile strength

D. Material Analysis

The pineapple leaf fiber (PALF) was examined and analyzed using a scanning electron microscope (SEM) and a light microscope. The results are summarized as follows:

(a) Bio-degradation Analysis: The SEM images (Fig. 10) presented clear microstructural differences between the samples before and after testing

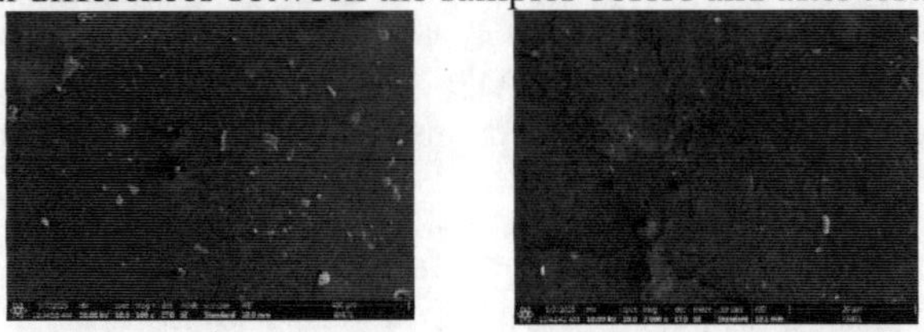

(a)

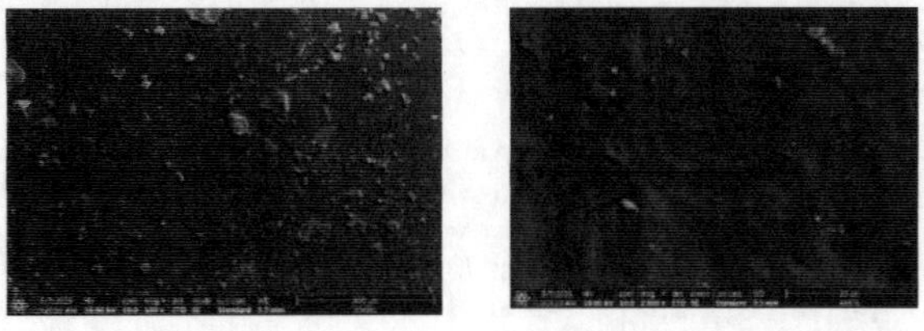

(b)

Fig. 10. Bio degradation (a: before testing (brown) and b: after testing (yellow)

- 100X magnification: The surface showed diffusion of fiber along with the presence of foreign objects. In addition, small cracks and uneven particles were observed surrounding the fiber surface (left).
- 10,000X magnification, the surface structure at the nanoscale appeared relatively smooth with small pores. Some interfacial bonding was observed between the fiber and the rubber matrix at the microstructural level (right)

(b) Water Absorption Analysis: The SEM images (Fig. 11.) revealed distinct microstructural changes in the samples when comparing their conditions before and after testing.

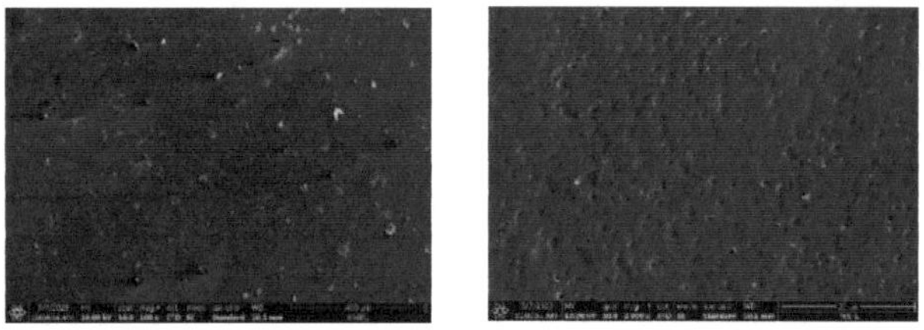

(a)

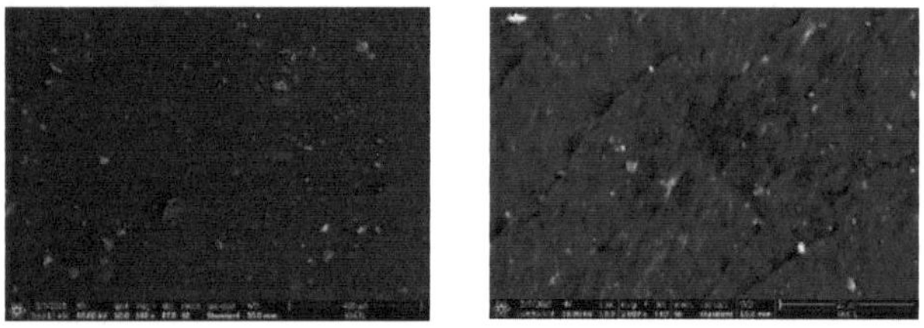

(b)

Fig. 11. Water absorption brown (a: before testing (brown) and b: after testing (yellow)

- 100X magnification, the surface appeared diffused and rugged. In addition, small scratches and cracks were observed at certain points.
- 10,000X magnification, numerous holes and cracks were observed distributed across the surface, indicating imperfections in the material at the microstructural level.

E. Optimization of response surface methodology for fabricating artificial leather combined with natural rubber

Based on the analysis of variance (ANOVA), it was found that X_1, X_2, and X_3 influenced the responses Y_1, Y_2, and Y_3. Of these, only Y_1 was further analyzed using Response Surface Methodology (RSM) because it was considered the most relevant for determining the optimal conditions calculated from the ANOVA for fabricating artificial leather combined with natural rubber.

In the surface plot (Fig. 12) used to identify the equilibrium value, it was observed that the maximum response for Y_1 (a) occurred in the relationship between X_1 and X_2. Specifically, when X_3 was fixed at 20 min, Y_1 showed the highest trend at an X_1 range of 4.5–5.0 g and X_2 between 90–100 °C. Furthermore, the X_1 and X_3 plot, with X_2 fixed at 90 °C, indicated a maximum Y_1 (b) for X_1

between 4.5–5.0 g and X_3 between 20–25 min. Finally, fixing X_1 at 5 g resulted in the highest Y_1 (c) response in the X_2 and X_3 plot, at X_2 between 90–100 °C and X_3 between 20–25 min. These combined observations suggest a highly favorable region for Y_1.

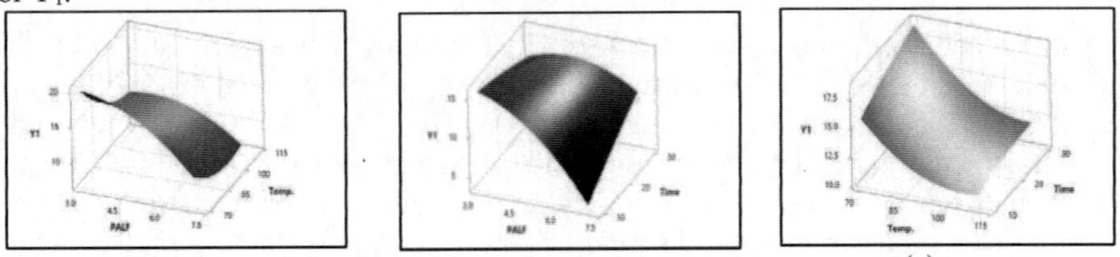

(a) (b) (c)

Fig. 12. Response Surface Plots illustrating the combined effects of X1, X2 and X3 on Y1

Conversely, the established model (Eq. 1) predicted a maximum biodegradability of 20.9946%. This optimal value was achieved under the following specific conditions (Fig. 11). achieved under the conditions of $X_1 = 3$ g, $X_2 = 70$ °C, and $X_3 = 10$ min, as shown in Fig. 13.

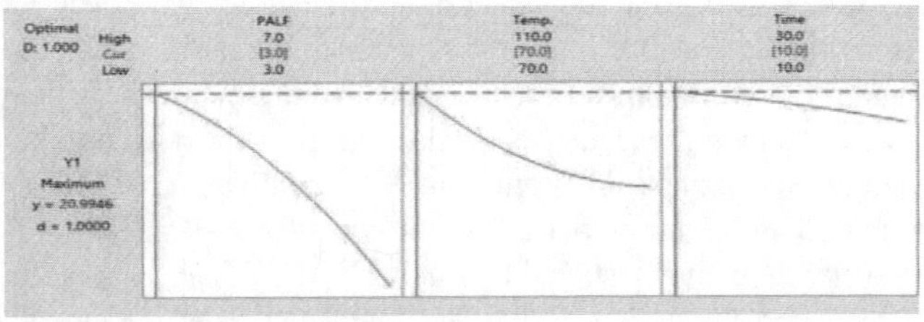

Fig. 13. Optimization of response surface methodology on the biodegradation

According to the model in Fig 11, the percentage of biodegradation tended to decrease as X_1 increased, which may be due to the higher fiber content making the material more resistant to degradation. In addition, increasing X2 and X3, which affect the compression of material, leads to higher density, potentially restricting access and thereby slowing the biodegradation process.

F. Perspectives on circular economy and sustainable development

Nowadays, the development of PALF from a traditional waste product into a valuable resource provides an excellent case study for the intersection of the circle economy and sustainable development [20-21]. While Thailand's pineapple cultivation has traditionally resulted in a significant amount of unused biomass [22-23], the emerging PALF industry demonstrates a paradigm shift from the linear take-make-waste model to a more regenerative, circular one [22]. This experiment demonstrates that the lifecycle of a PALF-based product can be extended beyond its initial use by recycling it into new materials at the end of its life, thereby reinforcing the circular model and reducing the need for virgin resources.

Crucially, this research quantifies resource efficiency in alignment with SDG 12 (Responsible Consumption and Production) [24]. The process begins with biomass valorization, where 1 kg of pineapple leaves yields only 20 g of viable PALF for experimental material. Furthermore, the RSM methodology ensures that comprehensive, multi-variable analysis is conducted with minimal material consumption: a total of only 75 g of PALF was required for the 15 experimental runs (approximately 3 kg of leaf pineapple leaves waste total), demonstrating the principles of Green Technology and efficient R&D [25-26]. This rigorous focus on minimizing material consumption supports sustainable industrial processes.

In addition, the development of PALF from Pineapple leaves to the achievement of some of the Sustainable Development Goals (SDG), particularly of SDG 7 (Affordable and Clean Energy), SDGs 9 (Industry, Innovation and Infrastructure) and SDG 12 (Responsible Consumption and Production) [27-28], through the provision of new bio-based products, development of novel and sustainable industrial processes, and boosting of regional and global economic development.

5 CONCLUSION

This research focused on synthetic polymers and single-source natural fibers, using Box– Behnken Design (BBD) and Response Surface Methodology (RSM) to study the effects of key factors (X_1, X_2, and X_3) on material properties (Y_1, Y_2, and Y_3). The results showed that X_1, X_2, and X_3 had a statistically significant effect (P-value) on Y_1, with the highest biodegradability (20.99%) obtained under the conditions of $X_1 = 3$ g, $X_2 = 70$ °C, and $X_3 = 10$ min. In contrast, Y_2 and Y_3 did not show statistically significant effects. However, although the factors did not significantly influence the P-values, the model still indicated trends: the minimum Y_2 was 10.90%, while Y_3 reached the highest value of 3.46 N/mm^2 under the conditions of $X_1 = 3$ g, $X_2 = 90$ °C, and $X_3 =$ 20 min. Nevertheless, no sustainable statistical significance was found in these responses. In contrast, synthetic polymers that have been developed demonstrate potential to support the circular economy model and reduce reliance on virgin resources from the agricultural sector in the future.

Author contributions

Sokjabok S. conceptualized the research idea and prepared the manuscript. Srisang S. revised the manuscript and provided suggestions regarding management. Hutangkoon T., Eiangmee O., and Maikaew T. conducted the experiments, carried out the evaluations, and investigated the system performance. Kunyuan J. and Srisang N. reviewed and approved the final version of the manuscript.

Funding

This work was financially supported by King Mongkut's Institute of Technology Ladkrabang [KREF186723]

References

[1] S. Chaudhary and B. Singh, "Pineapple by-products utilization: Progress towards the circular economy,"Food and Humanity, vol. 2, no. 1, p. 100243, 2024.
[2] M. Fadda, A. Zych, R. Carzino, A. Athanassiou, and G. Perotto, "Hydrophobic and water-resistant fish leather:A fully sustainable combination of discarded biomass and by-products of the food industry," Green Chemistry, vol. 26, no. 13, pp. 542–555, 2023.
[3] N. R. Ariyani, "Mycelium-based leather: Potential for sustainability: A review," AIP Conf. Proc., vol. 3001, p.030012, 2024.
[4] Y.-H. Choi and K.-H. Lee, "Ethical consumers' awareness of vegan materials: Focused on fake fur and fake leather," Sustainability, vol. 13, no. 1, p. 436, 2021.
[5] FAO, OECD-FAO Agricultural Outlook 2025–2034. Rome, Italy: Food and Agriculture Organization of the United Nations, 2025.

[6] A. F. A. Hamzah, M. H. Hamzah, H. Che Man, N. S. Jamali, S. I. Siajam, and M. H. Ismail, "Recent updates on the conversion of pineapple waste (Ananas comosus) to value-added products, future perspectives and challenges," Agronomy, vol. 11, no. 11, p. 2221, 2021.
[7] P. Usubharatana and H. Phungrassami, "Evaluation of opportunities to reduce the carbon footprint of fresh and canned pineapple processing in central Thailand," Pol. J. Environ. Stud., vol. 26, no. 4, pp. 1725–1735, 2017.
[8] S. Komsoon, A. Mongkholchai, and M. Panason, "Development and application of an automatic river-waste classification system in Surat Thani and Songkhla, Thailand," Int. J. Environ. Sci. Dev., vol. 16, no. 4, pp. 234– 242, 2025.
[9] N. T. Minh and H. N. Ngan, "Vegan leather: An eco-friendly material for sustainable fashion towards environmental awareness," AIP Conf. Proc., vol. 2406, p. 060019, n.d.
[10] E. K. Roh, "Mechanical properties and preferences of natural and artificial leathers, and their classification with a focus on leather for bags," J. Eng. Fibers Fabr., vol. 15, pp. 1–12, 2020.
[11] H. Katogi, "Effect of outdoor exposure on tensile property of synthetic leather for personal mobility,"Procedia Struct. Integr., pp. 611–617, 2024.
[12] J. Smith and K. Lee, "Initial mechanical properties of orange peel waste as raw material for vegan leather production," J. Sustain. Mater., vol. 15, no. 2, pp. 45–56, 2020.
[13] M. Muslimin, "Effect of processing treatment of pineapple leaf fiber (PLF) with liquid smoke on tensile properties and impact of pineapple fiber reinforced composites," Technium Science, vol. 17, no. 1, pp. 186– 189, 2023.
[14] A. Surname and B. Surname, "Development of green leather alternative from natural rubber and pineapple leaf fiber," J. Sustain. Mater., vol. 18, no. 2, pp. 101–112, 2023.
[15] S. Pichaiyut, C. Uttaro, K. Ritthikan, and C. Nakason, "Biodegradable thermoplastic natural rubber based on natural rubber and thermoplastic starch blends," J. Polym. Res., vol. 30, no. 1, p. 23, 2023.
[16] S. A. Riyajan, "Biodegradable green glove containing ascorbic acid from maleated epoxidized natural rubber/poly(vinyl alcohol) blend: Preparation and physical properties," J. Polym. Environ., vol. 30, no. 3, pp. 1141–1150, 2022.
[17] L. G. Oliveira, A. P. Paiva, P. P. Balestrassi, J. R. Ferreira, S. C. Costa, and P. H. Campos, "Response surface methodology for advanced manufacturing technology optimization: Theoretical fundamentals, practical guidelines, and survey literature review," Int. J. Adv. Manuf. Technol., vol. 104, pp. 1785–1837, 2019.
[18] Prasit, "The optimal conditions for the preparation of TiO_2 and polyamine thin films," M.S. thesis, Suranaree Univ. Technol., Nakhon Ratchasima, Thailand, 2019.
[19] Chaiwut, "Design of a separator and reclamation machine for latex," M.S. thesis, Kasetsart Univ., Bangkok, Thailand, 2017.
[20] P. K. Sarangi, A. K. Singh, R. K. Srivastava, and V. K. Gupta, "Recent progress and future perspectives for zero agriculture waste technologies: Pineapple waste as a case study," Sustainability, vol. 15, no. 4, p. 3575, 2023.
[21] L. I. S. Gonzatti, A Needle in a Pineapple Field: Study of the Current State of Pineapple Leaves Valorisation in the Context of Circular Bioeconomy in Costa Rica. 2023.
[22] K. Boonrod and P. Ninduangdee, "An opportunity on new biomass materials application for renewable energy in Thailand: Perspective of pineapple farmers," in Proc. 8th Int. Conf. Lang., Educ., Humanities Soc. Sci., Mar. 2018, pp. 27–29.
[23] B. O. Hemung, M. Sompholkrang, A. Wongchai, N. Chanshotikul, and N. Ueasin, "A study of the potential of by-products from pineapple processing in Thailand," Int. J. Health Sci., vol. 6, suppl. 4, pp. 12605–12615, 2022.
[24] United Nations in Thailand, "How the UN is supporting the Sustainable Development Goals in Thailand." [Online]. Available: https://thailand.un.org/en/sdgs. [Accessed: Sep. 24, 2025].
[25] P. Sharma, "Optimized conversion of waste vegetable oil to biofuel with meta-heuristic methods and design of experiments," J. Emerg. Sci. Eng., vol. 1, no. 1, pp. 22–28, 2023.
[26] S. Duangsuwan, P. Junkong, P. Phinyocheep, S. Thanawan, and T. Amornsakchai, "Development of green leather alternative from natural rubber and pineapple leaf fiber," Sustainability, vol. 15, no. 21, p. 15400, 2023.
[27] A. S. Sodhi, S. Bhatia, and N. Batra, "Advances in green technologies and global circular bioeconomy framework for a sustainable society," Circular Econ. Sustain., pp. 1–26, 2025.
[28] Mahidol University, "Pineapple leaf fibers for textiles and industry."[Online]. Available: https://sustainability.mahidol.ac.th/th/case-study/SDGs/detail/763. [Accessed: Sep. 24, 2025].

Performance Compatibility Assessment of Selected Motorcycles Using Nipa Bioethanol as Alternative Fuel Blend

Nathaniel Ericson Mateo[1,2*], *Aldrin Calderon*[2,3], *Shirley Agrupis*[4], *Kendrick Cyrus* Acidera[1], and *Loreli Faye Manzano*[4]

[1]ME Department College of Engineering, Mariano Marcos State University, 2906 Philippines
[2]School of Graduate Studies, Mapua University, Manila, 1002 Philippines
[3]School of Mechanical, Manufacturing, and Energy Engineering, Mapua University, Manila, 1002 Philippines
[4]National Bioenergy Research and Innovation Center, Mariano Marcos State University, 2906 Philippines

Abstract. The Philippines Biofuels Act of 2006 mandates 10% anhydrous (E10) ethanol blend in gasoline from E20 to E85 by 2030 but remains to E10. The potential of using hydrous (H) nipa-bioethanol remains unexplored which may alleviate production and volume requirements contributing to SDG7 Clean and Affordable Energy. Test vehicles must start and run normally to be concluded as fuel compatible. Findings show that hydrous blends were compatible from E20H to E40H, except on one carb motorcycle that experienced hard starting or unstable combustion during the test with E40H. The two carbureted motorcycles were observed to have an increased power developed from 11% to 29% using hydrous blends versus E10. Also, the two FI test motorcycles were observed to have about 3.5% to 8% power output improvements. However, the fuel consumption was observed to be less efficient with decreased distance travelled per liter by about 8% for every additional 10% increase of hydrous blend. Although hydrous blends are compatible with reduced HC and CO up to E40H, the study concludes that E20H is the optimal hydrous blend. Blend more than E40H requires remapping on FI while carb adjustments must be made on carb type motorcycles for blends greater than E30H.

1 Introduction

Fuel blends incorporating bioethanol have been widely adopted across the globe, driven by their potential to mitigate greenhouse gas emissions and safeguard the environment [1]. Given the well-established link between fossil fuels and climate change, bioethanol emerges as a pivotal renewable energy source, offering a pathway to reduce reliance on non-renewable resources [2], [3]. Significant effect of using fuel blends includes brake thermal efficiency improvement on engine performance, increase the gasoline fuel's Rating Octane Number or RON rating that improves anti-knocking properties on engines, and better smoke emissions

* Corresponding author: nrmateo@mmsu.edu.ph

© The Authors, published by EDP Sciences. This is an open access article distributed under the terms of the Creative Commons Attribution License 4.0 (https://creativecommons.org/licenses/by/4.0/).

[4], [5]. As regards health issues, bioethanol blends reduce carcinogen emissions in the atmosphere [6].

While many countries have mandated ethanol blends in gasoline, the Philippines has pursued its commitment under the Biofuels Act of 2006, requiring a 10% anhydrous ethanol blend, termed E10, with plans for an increase to E20 by 2020 and E85 by 2030 [7]. Despite aiming for a 20% ethanol utilization by 2020, the Philippines has remained at E10 due to challenges related to feedstock availability and limitations in installed plant capacity [8]. Similar concerns regarding ethanol scarcity for blending have been anticipated in other nations, including India and China [9], [10]. Despite these large-scale projects, the Philippines still imports the majority of its anhydrous bioethanol requirements. Producing anhydrous ethanol is expensive due to the high energy demand of the dehydration process, which accounts for about 37% of the total production cost [11]. The production of anhydrous bioethanol requires sophisticated equipment and large-scale plants. Meanwhile, the potential of using hydrous bioethanol remains underexplored [12]. As such, the potential of using hydrous bioethanol remains underexplored. Using aqueous ethanol (95% purity) as a fuel could improve overall energy efficiency, potentially saving 10-45% of energy by eliminating the need for dehydration [11].

Mariano Marcos State University (MMSU) in the Philippines, through its National Bioenergy Research and Innovation Center, has pioneered the adoption of hydrous ethanol (H) as a fuel blend. The center's study demonstrates that hydrous bioethanol, at a 20% blend, functions effectively as an anhydrous fuel blend. Hydrous ethanol as a fuel blend was also found to be promising in different studies coming up with better brake thermal efficiency, power, mileage, emissions [1], [13]. Fascinatingly, significant cost savings can be achieved by omitting the dehydration process when utilizing hydrous bioethanol in fuel blends [14]. Nevertheless, this approach presents challenges. Observations include difficulties with starting, occurrences of pitting and corrosion, and a decline in heating value [8].

MMSU aims to highlight the viability of employing a higher ratio of hydrous fuel blends for private use in its university vehicles, opting for this over anhydrous blend to capitalize on savings from bypassing the dehydration process in distillation facilities. Furthermore, MMSU plans to engage its community recipients in rural areas in the production of hydrous bioethanol [14]. However, certain studies propose the use or adaptation of fuel systems, particularly fuel injectors [15], and on Flex-Fuel vehicles [12], [16] suggest potential compatibility issues and room for improvement when employing fuel blends in gasoline engines.

This study conducted performance compatibility testing on selected motorcycles, utilizing hydrous bioethanol as a fuel blend, to determine the optimal ratio for effective operation without requiring any fuel setting adjustments or engine modifications.

2 Materials and Methods

The paper focused on the compatibility performance of nipa hydrous bioethanol as fuel blend to commercial gasoline E10 that may serve as basis in the implementation of hydrous fuel blend to pilot the utilization in MMSU that pioneered the production of fuel grade hydrous bioethanol in the Philippines showing the concept in Figure 1. The commercial E10 fuel serves as the base fuel blend to 95% azeotrope hydrous ethanol (H). The compatibility assessment on test motorcycles using E10 and various hydrous blends must operate without encountering any hard starting and combustion issues as indicated by the power output and fuel mileage without performing any adjustment on carb type test motorcycles or tuning/remapping on Fuel Injected (FI) test motorcycles used in this paper.

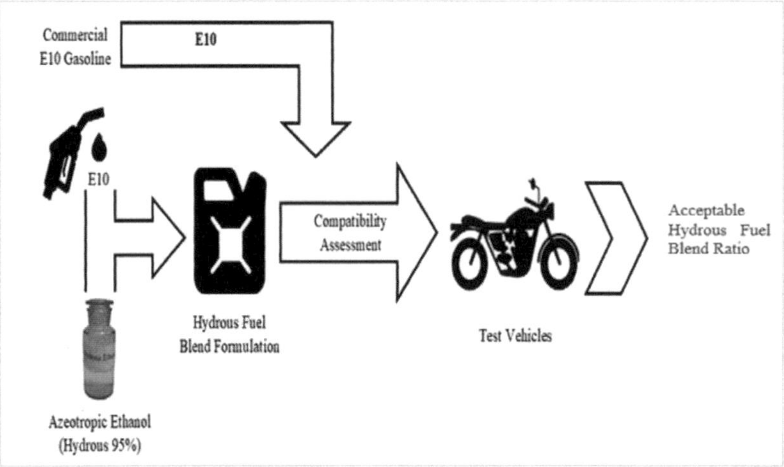

Fig.1. Conceptual Framework

2.1 Test Vehicles

The study evaluates the compatibility of hydrous gasoline fuel blends on four different motorcycles as test vehicles without any modifications to the engine and fuel system. Two test vehicles are carburetor-type (carb) engines, while the other two test vehicles are Fuel Injected (FI) engines as shown in Figure 2: Honda XR 125L, Suzuki Raider R150, Yamaha Sniper 155, and Honda TMX 155 from left to right respectively.

Fig.2. Test vehicle used in the study

The Honda TMX 155 and Honda XR 125 appeared to be the most and least used carbureted tricycles and service vehicles in the locale respectively, while the Suzuki Raider R150 and Yamaha Sniper 155 appeared to be the emerging favorite FI service motorcycles of students and enthusiasts in the area.

2.2 Hydrous Ethanol Fuel Blend Preparation and Formulation

Two brands of commercial gasoline with a 10% anhydrous fuel blend (E10) were used as control and base fuel: gasoline P and gasoline S. The selection was based on the greatest number of gasoline stations in Ilocos Norte in the Philippines, with 164 fuel stations having 69 different brands in the area. Gasoline S and gasoline P appeared to have the most significant number of commercial fuel stations and are available in the suburbs of the locale of the study. The hydrous fuel blends were formulated using the ratio of E10 to 95% azeotrope hydrous bioethanol as follows: 1:7.5 for E20H; 1:3.25 for E30H; and 1:1.83 for E40H.

The hydrous bioethanol fuel blend in this study came from Nipa Sap using MMSU-NBERIC's proprietary technology developed and used in previous studies [8]. Additional acid neutralization processes was made in the produced hydrous ethanol to address corrosion found in several studies [14], [17] and so as to be comparable with the Philippine National Standards (PNS) using ASTM D6423 for acidity/alkalinity, ASTM D4052 for density, ASTM D5501 for ethanol and methanol contents, and ASTM D1613 for total acetic acid determination.

2.3 Engine Performance Compatibility Testing

The compatibility test referred to in this study is the ability of the test motorcycles to adapt to the hydrous fuel blends without any hard starting issues and problems leading to combustion malfunction during testing. A motorcycle test dynamometer MT 511 rig was utilized in this study to measure the engine's performance in terms of power and air-fuel ratio based on the SAE 1349J procedure [13]. The kW power output of the MT 511 dynamometer refers to the water brake dynamometer output and not either the wheel power or engine brake power of the test vehicle. The rpm also refers to the dynamometer rpm, not the speed of the motorcycle. This is used primarily to compare the quantitative power output of the test vehicle using the hydrous blend formulations in comparison to E10 fuel.

A decibel meter or sound level meter was used to determine the sound level. The device is placed at a specific distance of 0.5 meters behind at the same level as the exhaust outlet, and at 45^0 angles to the motorcycle's direction of travel in compliance with the Measurement of Exhaust Sound Levels of Stationary Motorcycles (SAE J1287) [18].

The mileage run test is carried out around the main campus of MMSU to determine the fuel consumption simulating long-distance and city driving conditions based on the Japanese 1015 mode cycle procedure. The test run covered about 7.56 kilometers per trial, with three straight trial runs conducted simultaneously using the four test vehicles with the same fuel blend formulations.

Any sign of hard starting and undesired combustion process in a test vehicle using the fuel blends is an indicator implying that the said fuel blend ratio is no longer compatible to the said test vehicle during testing, both on dynamometer tests and on actual mileage runs.

3 Results and Discussions

3.1 Properties of Hydrous Bioethanol from Nipa Sap

Table 1 shows the properties of nipa hydrous ethanol in comparison to the Philippine National Standard (PNS) of an anhydrous ethanol. It shows that the 95% hydrous ethanol used in this study as a fuel blend is within the limits of anhydrous ethanol, except for the density, lower ethanol content, and higher heating value content [18] that is advantageous as fuels [5]. After the distillation of nipa feedstock, additional neutralization process was performed to reduce the total acetic acid of the hydrous ethanol used in this study.

Table 1. The nipa hydrous ethanol as compared to PNS standard.

Properties	PNS Standard	Results	Remarks
Appearance	Clear and Bright	Clear and Bright	Passed
Acidity/Alkalinity, pHe	6.5-9.0	7.2	Passed
Density at 20^{0}C, kg/m^3	791.5 max	809.8	Higher
Ethanol Content, %v/v	99.3 max	95.63	Lower
Methanol Content, %v/v	0.5 max	0.01	Passed
Total Acetic Acid), %v/v	0.007 max	0.0008	Passed
Heating Value, Btu/lb	11,484 [20]	11,707	Higher

3.2 Performance of the four-Test Motorcycles

Figure 3(a) displays the power-speed curve of the Suzuki Raider 150 while operating on fuel blends E10, E20H, E30H, and E40H, with gasoline P and gasoline S as the base fuel gasolines. The E20H produces the most power at 4.453 kW when blended with gasoline P. The highest recorded power produced is 5.238 kW using E10 fuel with gasoline S. The graph's trend indicates that power increases correspondingly with rpm.

However, ANOVA test results appear to be insignificant in terms of power between the various fuel blends in base gasoline P and S. The calculated Fstat value is 0.531712, which is lower than the Fcrit value of 2.08172 at 0.05 level of significance. The said test vehicle runs at normal conditions up to E40H.

In Figure 3(b) shown below depicts the performance of the Yamaha Sniper 155 in terms of power and speed curves. The E10 fuel blend from gasoline P is 5.97 kW, while the maximum recorded peak power from gasoline S is 6.25 kW. Despite E10 displaying the highest power produced, the said test vehicle runs at normal conditions up to E40H. The ANOVA results also show that there is no significant difference between the four fuel blends in the Yamaha Sniper155 at 0.05 level of significance with a computed Fstat value of 1.825397 which is less than the Fcrit value of 2.095504. Meanwhile, from the data shown in Figure 3(c), the ANOVA test shows that the computed F value of 0.274871 is lesser than that of the Fcrit value of 2.290432. This implies that there is no significant difference in power amongst the different fuel blends on the performance of the Honda XR125 when different fuel blends were used. E10 fuel blend produced the most power with 4.295 kW when the gasoline S was used. However, the Honda XR125 test vehicle found operable only up to E30H with gasoline S. Compatibility and performance tests failed at E40H due to hard starting conditions and unstable engine performance. The vehicle runs smoothly using E40H after some adjustments made in the carburetor setting. Figure 3(d) depicts the performance of the Honda TMX 155 with various fuel blends for both gasoline P and gasoline S. The highest power produced by an E10 fuel blend is 4.039 kW under gasoline P. The E10 fuel blend generates the most power in gasoline S which is 3.087 kW. In the ANOVA test, there appears to be no statistically significant difference in power output across the various fuel blends employed. At the 0.05 level of significance, the computed F value of 0.969 is less than the critical F value of 2.1065. The fuel blends work well up to E40H using both base fuels gasoline S and P.

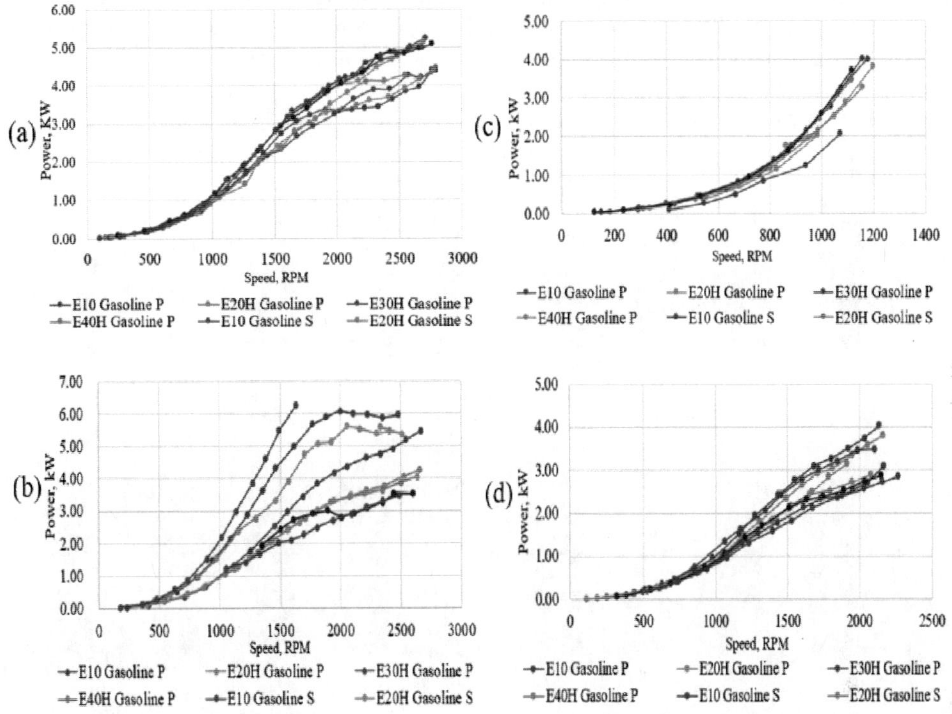

Fig. 3. (a) Power-speed curve of Suzuki Raider R150, (b) Power-speed curve of Yamaha Sniper 155, (c) Power-speed curve of Honda XR 125, (d) Power-speed curve of Honda TMX 155.

Most of the performance tests as to dynamometer waterpower output conducted on the four test vehicles appeared to have no significant difference when fueled with E20H to E40H when compared to E10 as the fuel of the test vehicles. This partially coincides with the results of other studies appearing to have a significant effect or at least comparable brake thermal efficiency [19]. Interestingly, the hydrous blend up to E20H compatibility on a sedan test vehicle [13] was surpassed by all four test motorcycle vehicles up to E40H except for Honda XR 125 which is compatible with E30H without any fuel setting calibrations or modifications.

The carburetor adjustment on Honda XR 125 to run smoothly on E40H suggests that hydrous blends of higher ratios are also possible as implied based on the flex-fuel article in Brazil which are already in the commercial stage and implies that modifications on the fuel systems of vehicles is required to adopt higher bioethanol gasoline fuel blends [16].

3.3 Mileage Test Run of the four-Test Motorcycles

As shown in Figure 4 are the fuel consumption per distance travelled of the four test vehicles implies that as the hydrous blend increases, the mileage runs also decrease using both hydrous blends with gasoline P and S as indicated by the decreasing bar trend. E10 fuel both from gasoline P and gasoline S offers the best fuel mileage of the four test vehicles used in this study. Some irregular mileage path was observed for both carbureted vehicles at E20H blend fuel with gasoline S and E30H with gasoline P on motorcycles Honda XR 125L and Honda TMX155 as shown in Figure (c) and Figure (d) below respectively. This is perhaps one of the disadvantages of a carbureted fuel system compared to FI systems known to be more efficient and quicker to respond [20].

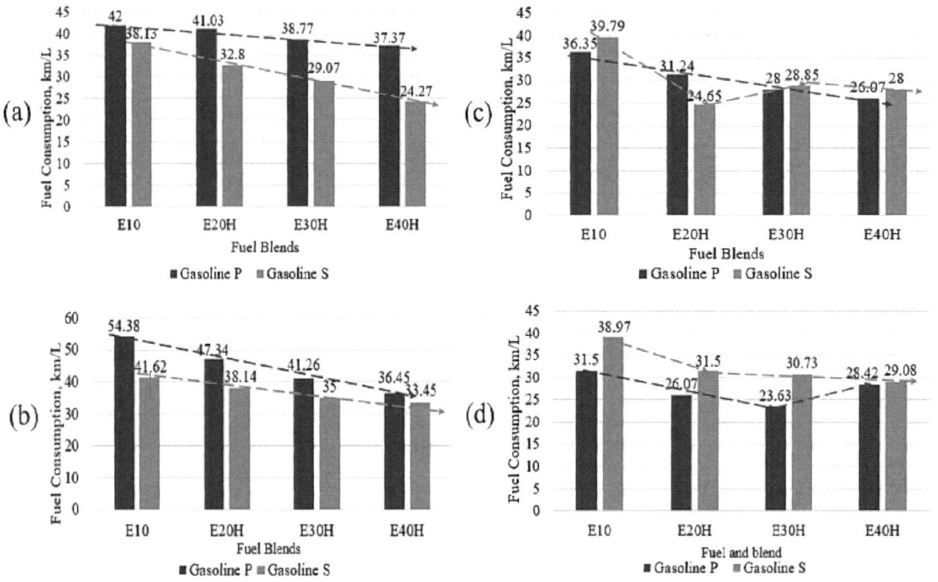

Fig.4. (a) Mileage Performance of Suzuki Raider R150, (b) Mileage Performance of Yamaha Sniper 155, (c) Mileage Performance of Honda XR 125, (d) Mileage Performance of Honda TMX 155.

The fuel consumption data on the Honda XR 125 in the mileage run was obtained upon some adjustments made in the carburetor using the E40H fuel blend with gasoline P which initially failed in the dynamometer test. This is to show that upon minimal setting adjustment, the said vehicle became compatible with E40H same as the other test vehicles. The Honda TMX 155 on the other hand was observed to be compatible up to E40H.

The fuel economy in this study was observed to decrease in FI test motorcycles by about 7% to 7.65%, while about 7.16% to 8.7% decrease in fuel economy for motorcycles with carbureted fuel system for every additional 10% hydrous fuel blend similar to a test of E20 decreasing by about 5% [21]. This is because of the diminishing heating value [18], [20] of the resulting blends as the ratio increases while maintaining the power output expected to deliver. A similar diminishing effect on performance was observed when an anhydrous blend was used instead [22].

However, this finding contrasted with a study by Mohammed et al that implies a higher energy input will lead to a better fuel economy [5]. Different results were also found on a sedan with decreased mileage while an increased mileage was found on a motorcycle using E20H when compared to E10 based on percentage difference [14].

Despite the different findings as to the fuel economy of the test vehicles, hydrous fuel blends up to E40H still suggest that the said hydrous fuel blends compatibly will function well on the different vehicles either in carbureted or FI vehicles used in this study.

3.4 Dynamometer Performance of the Four Test Vehicles

Very high and very low A/F ratios may cause either fuel-efficient or incomplete combustion of fuel due to either less fuel or much fuel present in the combustion chamber respectively [23]. The power developed for all hydrous blends on test vehicles was observed at the ideal A/F equivalent of 14.7 [24] to compare the performance of the said fuel blends. Figure 5 below shows both carbureted testing vehicles Honda TMX 155, and Honda XR 125L generate the highest power under E20H with gasoline P with 3.578kW and 4.283kW power output respectively.

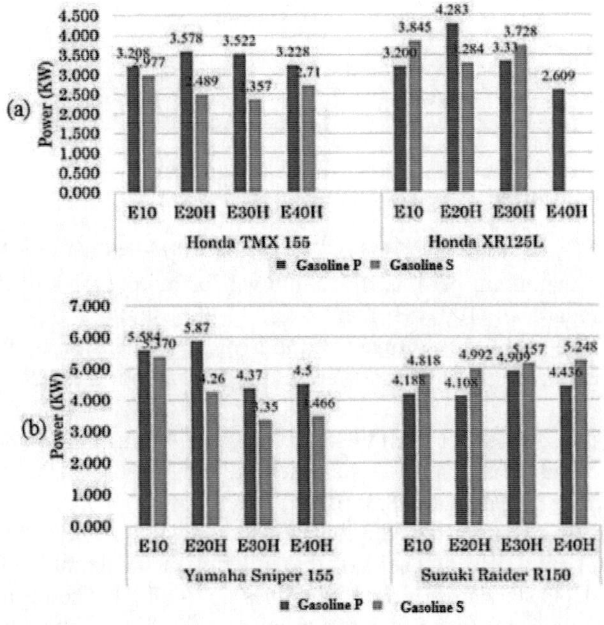

Fig. 5. (a) Power generated of carbureted motor vehicle at ideal A/F ratio (b) Power generated of FI motor vehicle at ideal A/F ratio.

Also shown in Figure 5 above is the power developed performance of the two FI test vehicles. The Yamaha Sniper performs better using E20H with gasoline P versus E10 with 5.87kW power output. The Suzuki Rider shows better performance as the hydrous blend increases when compared to E10 and best performs in E30H with gasoline P with 4.909kW power output.

Tests showed improved power performance of E20H with gasoline P when compared to E10 as fuel on Honda TMX 155 by 10.9%, on Honda XR 125L by 28.9% and on Yamaha Sniper by about 5%. The Suzuki Rider improved by 15.9% with E30H with gasoline P and by 8.5% with E40H with gasoline S versus E10.

The power developed at ideal A/F shows that the hydrous fuel blend produces higher power more often using E20H in the three test vehicles when compared to E30H and E40H. Despite the observed decrease of power developed at ideal A/F [21], the test vehicles extends performance compatibility up to E40H blend except to the Honda XR 125L up to E30H.

However, this incompatibility was addressed after adjustments made in the carb jet of the said test vehicle.

3.5 Emission Performance

As expected, hydrous blend found to be environmentally friendly based on the test conducted on one of the test vehicles used in this study, the Suzuki Raider FI motorcycle as shown below in table 2. Improved emission on the test motorcycle shows about 8% and 96% decrease in CO and HC emissions respectively. This is supported by the 4% increase in the CO2 emissions from E10 to E30H, an indication of a better combustion inside the engine cylinder [25].

Table 2. Emission Performance of Raider Test Vehicle

Gas	E10	E20	E30
CO	7.13%	6.46%	6.58%
CO2	6.44%	5.4%	6.2%
HC	848 ppm	323 ppm	296 ppm

3.6 Emission Performance

The Philippine Land Transportation Office (LTO) issued Administrative Order No. 2010-32, which states that the maximum permitted sound level for motorcycles is 99 decibels (dB) lower than the standard of 100 dB mentioned in the study about muffler silencer modifications [26]. The said limit is considered to avoid apprehension by law enforcement agencies when the said hydrous fuel blends are utilized for piloting either inside or outside the university.

In this research, the sound level from the four test vehicles ranges from 84.7 dB to 98 dB when using gasoline P as a fuel blend. Also, when utilizing gasoline S as a fuel blend, the sound level ranges from 85.2 dBA to 98.5 dBA.

All the sound levels gathered from the four test motorcycles are less than the maximum permitted sound level of 99 dBA and suggest that utilizing the different hydrous fuel blends is acceptable by the LTO standards of the Philippines up to E40H. The result also coincides partially with the lowering noise effect found when E10H and E20H were used when compared to neat gasoline [27].

4 Conclusion and Recommendations

In the dynamometer test conducted with four different motorcycles, hydrous fuel blends exhibited performance comparable to E10, with no significant differences detected. Compatibility was found across blends from E20H to E40H, except for the Honda XR 125L, which experienced issues with the E40H blend with gasoline S. However, minor adjustments to the carburettor resolved this, ensuring compatibility similar to the other motorcycles.

In terms of air/fuel mixture optimization, crucial for maximizing performance, economy, and preventing engine damage, ANOVA analysis indicated a significant variance among the four test motorcycles. The computed F value of 8.352078 exceeded the critical F value of 2.960351, signifying significance at $p < 0.05$. While insignificant in output power performance, this suggests varying engine chamber performance with the use of hydrous blends across different vehicles.

Moreover, the power developed was analysed at the ideal A/F ratio for each motorcycle, and percentage differences compared to E10 were observed. Two carburetted motorcycles demonstrated improved power development with the E20H blend with gasoline P, ranging from 11% to 29%, but showed reduced performance with gasoline S blend. The FI test motorcycle, based on the Suzuki Raider, showcased better power development with hydrous blends, ranging from 15% to 6% for E30H and E40H compared to E10, respectively, using gasoline P, and 3.5% to 8.5% from E20H to E40H, respectively with gasoline S blend. The Yamaha Sniper 155 FI test motorcycle displayed diminishing power development effects with all hydrous blends except E20H, showing a 5% improvement in power developed compared to E10 using gasoline P. Emission on one of the FI motorcycles show about 8% and 96% decrease in CO and HC emissions respectively that is an indication of a more environment friendly blend due to the 4% increase in CO2 emissions from E10 to E30H, also an indication of a better combustion inside the engine cylinder.

Sound performance tests using hydrous blends up to E40H met LTO requirements, ensuring compliance without vehicle modifications. However, observed fuel economy decreased by approximately 7% to 7.65% for FI motorcycles and 7.16% to 8.7% for motorcycles with carbureted fuel systems for every additional 10% hydrous fuel blend, suggesting potential economic implications with higher blend ratios.

Therefore, utilizing hydrous blends up to E30H for carbureted motorcycles and up to E40H for FI motorcycles can be considered, as they passed compatibility performance tests. However, opting for the E20H blend may offer better mileage compared to the E30H and E40H for motorcycles.

References

1. X. Wang *et al.*, "Evaluation of hydrous ethanol as a fuel for internal combustion engines: A review," *Renew. Energy*, vol. 194, pp. 504–525, July 2022.
2. T. J. Tse, D. J. Wiens, and M. J. T. Reaney, "Production of bioethanol—A review of factors affecting ethanol yield," *Fermentation*, vol. 7, no. 4, p. 268, Nov. 2021.
3. A. I. Osman, L. Chen, and M. Yang, "Cost, environmental impact, and resilience of renewable energy under a changing climate: a review," *Environ Chem Lett*, vol. 21, pp. 741–764, 2023.
4. J. E. Anderson and T. J. Wallington, "Novel Method to Estimate the Octane Ratings of Ethanol-Gasoline Mixtures using Base Fuel Properties," *Energy & Fuels*, 2020.
5. M. K. Mohammed, H. H. Balla, Z. M. H. Al-Dulaimi, Z. S. Kareem, and M. S. Al-Zuhairy, "Effect of ethanol-gasoline blends on SI engine performance and emissions," *Case Stud. Therm. Eng.*, vol. 25, no. 100891, p. 100891, June 2021.
6. S. Mueller, G. Dennison, and S. Liu, "An assessment on ethanol-blended gasoline/diesel fuels on cancer risk and mortality," *Int. J. Environ. Res. Public Health*, vol. 18, no. 13, p. 6930, June 2021.
7. F. M. S. M. Haas, *Annual Biofuels Report," United States Department of Agriculture, Foreign Agricultural Service*. 2021.
8. N. E. R. Mateo *et al.*, "Nipa Sap (Nypa fruitican) as Supplementary Feedstock for Bioethanol Production in the Philippines: The Pamplona Experience," in *Springer Proceedings in Earth and Environmental Sciences*, Cham: Springer International Publishing, 2023, pp. 19–28.
9. B. Dey, B. Roy, S. Datta, and T. S. Ustun, "Forecasting ethanol demand in India to meet future blending targets: A comparison of ARIMA and various regression models," *Energy Rep.*, vol. 9, pp. 411–418, Mar. 2023.

10. A. Devi *et al.*, "A panoramic view of technological landscape for bioethanol production from various generations of feedstocks," *Bioengineered*, vol. 14, no. 1, pp. 81–112, Dec. 2023.
11. Z. Z. I. K. Tola E Al-Gaadi Alameen and A. A. Zeyada, "Hydrous Ethanol-gasoline Blends as Alternative Fuels for Spark Ignition Engine: Fuel Properties and Engine Performance," *International Energy Journal*, pp. 245–254, 2022.
12. A. Belonio, Philippine Rice Research Institute (PhilRice), Science City of Munoz, Nueva Ecija, Philippines, M. J. Regalado, N. Tado, and E. Sicat, "Development of a hydrous ethanol fuel feeding device for spark- ignition engine," *J. Technol. Innov. Renew. Energy*, vol. 3, no. 4, pp. 159–165, Dec. 2014.
13. J. L. Yu and E. N. Quiros, "Performance characteristics of Philippine hydrous ethanol-gasoline blends: Preliminary findings," in *ASME 2019 13th International Conference on Energy Sustainability*, Bellevue, Washington, USA, 2019.
14. N. E. R. Mateo, R. A. Ulep, M. P. Lucas, S. C. Agrupis, J. Sagadraca, and C. Baga, "An overview on the viability of hydrous bioethanol as gasoline fuel blend in the Philippines," in *The AUN/SEED-Net Joint Regional Conference in Transportation, Energy, and Mechanical Manufacturing Engineering*, Singapore: Springer Nature Singapore, 2022, pp. 301–314.
15. P. P. Paenpong, "Development of electronic fuel Injection system through the use of Arduino Mega 2560 microcontroller," *Journal of Engineering and Innovation*, pp. 34–43, 2023.
16. T. C. C. De Melo, "Hydrous ethanol-gasoline blends - Combustion and emission investigations on a Flex-Fuel engine," *Fuel (Lond.)*, vol. 97, pp. 796–804, 2012.
17. L. Matějovský, J. Macák, M. Pospíšil, P. Baroš, M. Staš, and A. Krausová, "Study of corrosion of metallic materials in ethanol–gasoline blends: Application of electrochemical methods," *Energy Fuels*, vol. 31, no. 10, pp. 10880–10889, Oct. 2017.
18. Muhaji and D. H. Sutjahjo, "The characteristics of bioethanol fuel made of vegetable raw materials," *IOP Conf. Ser. Mater. Sci. Eng.*, vol. 296, p. 012019, Jan. 2018.
19. V. Edwin Geo, D. Jesu Godwin, S. Thiyagarajan, C. G. Saravanan, and F. Aloui, "Effect of higher and lower order alcohol blending with gasoline on performance, emission and combustion characteristics of SI engine," *Fuel (Lond.)*, vol. 256, no. 115806, p. 115806, Nov. 2019.
20. N. Sutarna, P. N. Bali, I. N. L. Antara, and D. S. Anakottapary, "Fuel consumption analysis of injection system and carburetor system on Honda beat fi 2013," *Logic*, vol. 20, no. 3, pp. 141–144, Nov. 2020.
21. A. Yakın and R. Behçet, "Effect of different types of fuels tested in a gasoline engine on engine performance and emissions," *Int. J. Hydrogen Energy*, vol. 46, no. 66, pp. 33325–33338, Sept. 2021.
22. D. Y. Dhande, N. Sinaga, and K. B. Dahe, "Study on combustion, performance and exhaust emissions of bioethanol-gasoline blended spark ignition engine," *Heliyon*, vol. 7, no. 3, p. e06380, Mar. 2021.
23. O. C. Nwufo, O. M. I. Nwafor, and J. O. Igbokwe, "Effects of blends on the physical properties of bioethanol produced from selected Nigerian crops," *Int. J. Ambient Energy*, vol. 37, no. 1, pp. 10–15, Jan. 2016.
24. A. Demirbas and K. Al-Ghamdi, "Relationships between specific gravities and higher heating values of petroleum components," *Pet. Sci. Technol.*, vol. 33, no. 6, pp. 732–740, Mar. 2015.

25. I. Made Suarta, I. Nyoman Gede Baliarta, I. P. G. Sopan Rahtika, and P. W. Sunu, "The role of hydrogen bonds of the azeotropic hydrous ethanol fuel composition to the exhaust emissions," *J. Phys. Conf. Ser.*, vol. 953, p. 012070, Jan. 2018.
26. O. A. Odunlami, O. K. Oderinde, F. A. Akeredolu, J. A. Sonibare, O. R. Obanla, and M. E. Ojewumi, "The effect of air-fuel ratio on tailpipe exhaust emission of motorcycles," *Fuel Communications*, vol. 11, no. 100040, p. 100040, June 2022.
27. S. A. Flamarz Al-Arkawazi, "The gasoline fuel quality impact on fuel consumption, air-fuel ratio (AFR), lambda (λ) and exhaust emissions of gasoline-fueled vehicles," *Cogent Eng.*, vol. 6, no. 1, Jan. 2019.

Characterization and Performance of Pineapple Stem Cationic Starch as an Eco-Friendly Flocculant for Phosphate Removal from Wastewater

Eden S. Erasga[1,*], *Angela Mae A.* Llames[2], *Kim Z.* Sangab [2], and *Angelica Nala T.* Yap[2]

[1]College of Arts and Science, Mapua Malayan Colleges Laguna, Cabuyao City, Philippines
[2]Department of Chemical Engineering, Mapua Malayan Colleges Laguna, Cabuyao City

Abstract. Municipal wastewater contains high levels of phosphate, which contributes to eutrophication in aquatic ecosystems. While coagulation-flocculation is a widely used method for phosphate removal, conventional coagulants often present challenges related to environmental sensitivity and performance under varying pH. This study investigates the use of a starch-based cationic coagulant derived from pineapple stem waste for phosphate removal in simulated municipal wastewater. The starch was chemically modified through etherification with glycidyltrimethylammonium chloride (GTAC), introducing quaternary ammonium groups to improve coagulation efficiency. Structural modifications were confirmed using Fourier Transform Infrared Spectroscopy (FTIR), which showed successful substitution of hydroxyl groups with ammonium functionalities. Coagulation-flocculation experiments were conducted to evaluate the effects of pH and coagulant dosage on phosphate removal. The modified starch achieved the highest phosphate removal efficiency at a dosage of 1.0 g/L and pH 5, significantly outperforming unmodified starch. The results demonstrate that chemical modification enhanced the starch's performance as a coagulant under acidic conditions. This study highlights the potential of utilizing modified starch from agricultural waste as an effective and sustainable alternative for phosphate removal, offering a promising approach for improving wastewater treatment technologies.

1 Introduction

Phosphorus (P) is one of the most essential macronutrients for plant growth and development, ranking second in importance only to nitrogen [1]. In the form of phosphate (PO_4^{3-}), it is extensively applied in agricultural fertilizers and animal supplements [2]. However, the increasing use of phosphates has resulted in widespread contamination of water resources. The primary contributors to this contamination are: (a) agricultural runoff, where excess phosphate fertilizers dissolve in rainwater and enter surface waters, and (b)municipal sewage,

* Corresponding author: eserasga@mcl.edu.ph

© The Authors, published by EDP Sciences. This is an open access article distributed under the terms of the Creative Commons Attribution License 4.0 (https://creativecommons.org/licenses/by/4.0/).

where phosphates persist in treated sludge and effluent [3].

The extensive release of bioavailable phosphates from agricultural and urban sources has accelerated eutrophication, contributing to the degradation of freshwater ecosystems [4]. Eutrophication is characterized by nutrient enrichment, particularly nitrogen and phosphorus, which promotes excessive algal and aquatic plant growth. This process reduces dissolved oxygen, increases turbidity, and degrades water quality, posing ecological and public health risks [5]. Since phosphorus is often the limiting nutrient in freshwater systems, even minor increases can trigger significant ecological shifts [6]. In response, various regulations have been implemented globally. For instance, the European Union sets discharge limits ranging from <1 to 2 mg P/L, depending on population equivalence [7]. In the Philippines, the Department of Environment and Natural Resources (DENR) Administrative Order No. 2016-08 mandates a maximum phosphate concentration of 0.025 ppm as phosphorus for Class C waters. Eutrophication events, such as algal blooms or red tides, commonly occur when total phosphorus (TP) exceeds 0.03 ppm [2]. Total phosphorus levels in municipal effluent commonly range from 6 to 8 mg/L [5].

Conventional phosphate removal technologies include biological treatment and chemical precipitation [4]. While biological treatment is highly efficient, it is associated with high operational costs, energy consumption, and land requirements, in addition to the need for precise aeration control [5]. Chemical precipitation offers a cost-effective alternative but often generates large volumes of sludge, poses pH control challenges, and can lead to secondary pollution [8].

Coagulation-flocculation has emerged as a viable treatment technology due to its operational simplicity and effectiveness [9]. The process consists of coagulation, flocculation, and sedimentation. Coagulation involves charge neutralization of colloidal particles using metal salts such as aluminum, iron, or calcium-based coagulants [5,10]. This is followed by flocculation, a slower, physical process wherein destabilized particles aggregate into larger flocs with the aid of organic polymers that act through bridging and charge reduction mechanisms [10,11]. These agglomerates then settle via gravity during sedimentation [12].

Due to growing concerns over the environmental and health impacts of conventional chemical coagulants—including $Al_2(SO_4)_3$, $Fe_2(SO_4)_3$, $AlCl_3$, and $FeCl_3$—there has been increasing interest in natural alternatives [9]. Natural coagulants offer several advantages, including biodegradability, renewability, nontoxicity, and lower sludge production [13]. These are typically polymeric substances such as microbial polysaccharides, starch, gelatin, galactomannans, cellulose derivatives, chitosan, glues, and alginate [14]. Their coagulation activity is attributed to the presence of functional groups such as hydroxyl (–OH), carboxyl (–COOH), and amino ($-NH_2$) groups, which can interact with suspended particles [13]. However, native forms of these biopolymers often exhibit limitations such as low water solubility, narrow effective pH ranges, and moderate charge density, which restrict their practical application [12]. Chemical modifications—such as esterification, etherification, and graft copolymerization—have been shown to enhance their coagulation performance by improving solubility, stability, and charge characteristics [15].

Starch-based coagulants, in particular, have garnered attention due to their structural abundance of hydroxyl and carboxyl groups on the polysaccharide backbone, making them amenable to modification [16]. Starches are commonly extracted from tubers and cereals, but are also found in high quantities in fruit-processing waste, such as skins, seeds, stems, and leaves [17]. Starch directly obtained from plant material is referred to as native starch, whereas chemically, physically, or biologically treated starches are termed modified starches. Pineapple (*Ananas comosus*) stems, for example, contain up to 77.78% starch by dry mass and have a purity of 97.77%, making them a promising raw material for value-added applications [18,19]. With global pineapple cultivation spanning over 900,000 hectares and the Philippines ranking as the second-largest exporter in 2023, producing 611,873 tonnes [20], pineapple stem starch represents a sustainable and abundant feedstock.

Although native starch can induce coagulation through charge neutralization and bridging mechanisms, it often suffers from low coagulation efficiency due to limited water solubility and low surface charge [12]. To overcome this, starches are commonly modified to form cationic derivatives that incorporate positively charged groups such as amino, imino, or ammonium, enhancing their ability to destabilize negatively charged colloidal particles [16,20]. Cationic starch coagulants have shown improved phosphate removal by enhancing charge neutralization and floc formation across a broad pH range [12].

This study investigates the phosphate removal performance of a cationic starch coagulant derived from pineapple stem starch, chemically modified via etherification with glycidyltrimethylammonium chloride (GTAC). The modified and unmodified starches were characterized using Fourier Transform Infrared Spectroscopy (FTIR), and their phosphate removal efficiencies were evaluated in simulated municipal wastewater across varying pH conditions and coagulant dosages. This work addresses the growing need for sustainable, bio-based coagulants by exploring the potential of agricultural waste valorization within circular wastewater treatment systems.

2 Materials and methods

2.1 Extraction of Starch from Waste Pineapple Stems

The pineapple stems were collected, washed with tap water, and rinsed with distilled water. Then, stems were peeled and cut into approximately 10 cm cubes. The cubes were blended with distilled water at a 1:2 weight ratio of pineapple stems to water, resulting in a fibrous paste. This slurry was filtered through cheesecloth to separate the starch-containing filtrate. The residue was discarded, and the filtrate, which contained both starch and water, was allowed to settle until a clear separation between the liquid fraction (supernatant) and the starch fraction (sludge) was observed. The liquid fraction was decanted and discarded. The starch sludge was air-dried for 1 d and then dried in a convection oven at 60 °C until a constant weight was achieved. The dried starch was sieved through a 100-mesh screen and stored at room temperature in an airtight jar with silica gel as a desiccant.

2.2 Synthesis of Cationic Starch Coagulant by Etherification.

In a 50 mL beaker, 5 g of pineapple stem starch were combined with 25 mL of 2 M NaOH. The mixture was stirred thoroughly at 50 °C for 1 h. Next, a cationic etherifying agent solution was prepared by dissolving 18.86 g of glycidyltrimethylammonium chloride (GTAC) in 10 mL of deionized water. Then, 6 mL of an 80 % GTAC solution in water was added to the starch mixture. The starch paste was stirred thoroughly for 6 h with continuous heating.

The reaction was stopped by adding 25 mL of deionized water, stirring without heat, and allowing the mixture to cool to room temperature. The mixture was then placed in an ice-water bath while the solution was neutralized to pH 7 using 1 M HCl. 95 % ethanol was added dropwise until white precipitates no longer formed, indicating that the reaction was complete.

The solution was decanted to remove the clarified ethanol layer on top. The remaining precipitate was obtained by filtration then it was rinsed thoroughly with 70 % ethanol to remove any unreacted reagents. The starch slurry was dried in a conventional oven at 50 °C until constant weight. The dried starch was pulverized using a mortar and pestle, then sieved through a 100-mesh screen. Finally, the modified starch was stored at room temperature in an airtight jar with a silica gel desiccant.

2.3 Characterization of Unmodified Starch and Cationic Starch

The unmodified and cationic starches were analyzed using Fourier Transform Infrared (FTIR) Spectroscopy to identify and compare their functional groups.

2.4 Preparation of Simulated Municipal Wastewater

A 200-ppm stock solution of potassium dihydrogen phosphate was prepared using distilled water. Simple dilution of the stock solution was performed to prepare the 10.0 ppm working solutions. The initial pH was measured using pH meter and the actual phosphate concentration of the solutions was measured using Phosphate High-Range Checker® HC-HI717 by HANNA Instruments.

2.5 Coagulation-Flocculation Experiments

A 3 % (w/v) stock solution was prepared for both unmodified and cationic starch by dissolving 3 g of starch in 100 mL of distilled water. To obtain starch concentrations of 1.0, 1.5, 2.0, 2.5, and 3.0 g/L, simple dilution of the stock solution was carried out.
To investigate the effect of pH on the phosphate removal efficiency of unmodified and cationic starch, batch tests were conducted using five 250 mL beakers, each containing 100 mL of 10 ppm simulated municipal wastewater. The pH of each beaker was adjusted to 5, 6, 7, 8, and 9 by adding 0.1 M NaOH or 0.1 M HCl dropwise as needed. Coagulation was then initiated by adding 9 mL of a 1.0 g/L unmodified starch solution to the wastewater, followed by rapid stirring at 295 rpm for 3 min. Flocculation was carried out by reducing the stirring rate to 25 rpm for 20 min. After stirring, the flocs were allowed to settle until a clear separation between the flocs and the liquid above was visible. The settling time was recorded, and the final pH and phosphate concentration were measured using a pH checker and a high-range (HR) phosphate checker, respectively. The same procedure was repeated using a cationic starch solution for comparison
The effect of coagulant dosage on phosphate removal was analyzed using the pH that resulted in the highest removal efficiency from the previous experiment. The same procedure was followed, with the exception that the coagulant dosage was varied at 1.0, 1.5, 2.0, 2.5, and 3.0 g/L for both unmodified starch and cationic starch. After the coagulation-flocculation process, the final pH and phosphate concentration were measured.

3 Results and discussion

3.1 Characterization of Unmodified Starch and Cationic Starch

The FTIR spectra of both unmodified and cationic starch samples are illustrated in Fig. 1. The spectra reveal a strong and broad absorption band at 3294.33 cm^{-1}, indicative of the stretching vibrations of hydroxyl (O-H) groups on the surface of unmodified starch. A peak at 2930.66 cm^{-1} corresponds to the stretching vibrations of sp^3 hybridized carbon-hydrogen (C-H) bonds, while the peak at 1642.54 cm^{-1} is attributed to the bending vibrations of O-H groups in tightly bound water, reflecting the hygroscopic nature of starch. Additionally, an absorption peak at 1149.72 cm^{-1} is associated with the stretching vibrations of aliphatic ether (C-O-C) groups, which is characteristic of the starch backbone. In contrast, the FTIR spectrum of the cationic starch displays an additional band at 1433.45 cm^{-1}, attributed to the stretching vibrations of quaternary ammonium groups (C-N). This peak confirms the successful incorporation of the GTAC cationic moiety onto the starch backbone.

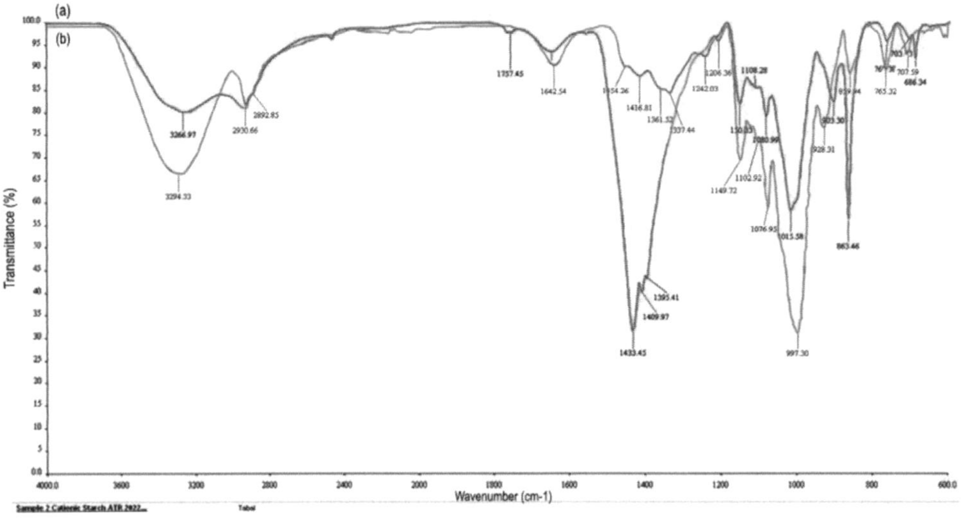

Fig. 1. IR spectra of (a) cationic starch and (b) unmodified starch.

3.2 Effects of Adsorption Operating Parameters

3.2.1 Effect of pH

The natural pH of untreated municipal wastewater is around 6.8, with phosphorus predominantly existing as dihydrogen phosphate ($H_2PO_4^-$) at a pH of 5. To optimize phosphate removal near the natural pH, the study recommended an initial pH of 5 for the coagulation-flocculation process, with treatment conditions including a 1.0 g/L coagulant dosage, 3 min of slow stirring, 20 min of fast stirring, and 1 h of settling time.

At 95 % confidence level, phosphate removal with unmodified starch significantly decreases beyond pH 5, with the highest removal rate of 40.34 % achieved at pH 5, as shown in Fig. 2 (left). The high phosphate removal rate at low pH is attributed to improved neutralization of negatively charged phosphate particles and protonation of negatively charged groups on the coagulant, enhancing destabilization and flocculation through hydrogen bonding. Whereas, under basic conditions, the hydroxyl groups of the starch molecule become negatively charged due to deprotonation, resulting in electrostatic repulsion with the negatively charged monohydrogen phosphate (HPO_4^{2-}) and thereby reducing phosphate removal efficiency. Unmodified starch, as a non-ionic polymer, reduces phosphate concentration in simulated municipal wastewater primarily through hydrogen bonding and polymer bridging, where longer polymer chains enhance the bridging process between particles to form larger settling flocs.

Results for cationic starch, as depicted in Fig. 2 (right), showed that the highest percent removal is 63.24 % at pH 5 and phosphate removal decreases as pH increases. At pH 5, where https://doi.org/50.50(5/e3sconf/10166900100(

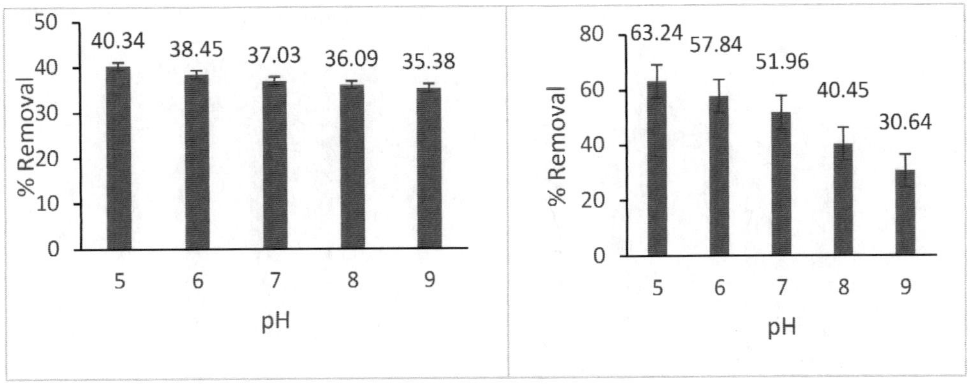

Fig. 2. Effect of pH on phosphate removal using unmodified starch (left) and cationic starch (right).

phosphate primarily exists as dihydrogen phosphate ($H_2PO_4^-$), the cationic starch modified with glycidyltrimethylammonium chloride (GTAC) is particularly effective for phosphate removal due to electrostatic attraction between the negatively charged phosphate species and the permanently charged quaternary ammonium groups introduced by GTAC. Many cationic polymers, particularly quaternary amines, carry a permanent positive charge and are thus largely unaffected by pH variations in suspension [21]. The cationic quaternary ammonium salt groups significantly increased the surface charge of starch from negative to positive [12]. This highly positive charge is advantageous for coagulants, as it effectively interacts with the negatively charged phosphate particles. The reduced effectiveness at pH 9 may be attributed to electrostatic repulsion, as deprotonated hydroxyl groups on cationic starch acquire negative charges that repel the negatively charged HPO_4^{2-}, leading to weaker interactions and lower removal efficiency. The modified starch was found to be pH-sensitive, as its charge neutralization effect diminished with increasing pH due to reduced surface charges and the collapse of its macromolecular structure [16].

3.2.2 Effect of coagulant dosage

The coagulation dosage is crucial for optimal coagulant performance in wastewater treatment, as both insufficient and excessive amounts can compromise impurity removal efficiency. Therefore, identifying the ideal dosage is essential for minimizing costs, reducing sludge formation, and ensuring effective treatment. The impact of coagulant dosage on phosphate removal was studied using dosages ranging from 1 g/L to 3 g/L at pH 5. The experiment involved treating 100 mL of a 10 ppm phosphate solution, with initial stirring at 295 rpm for 3 min, followed by a slower stirring at 25 rpm for 20 min at room temperature.

As shown in Fig. 3 (left), the average phosphate removal percentage increases with higher coagulant dosages, peaking at 39.86 % with a 3.0 g/L dose and dropping to 34.20 % with a 1.0 g/L dose. The data also indicate only minimal improvement in removal efficiency when increasing the dosage from 2.0 to 3.0 g/L. Higher coagulant dosages enhance phosphate removal by improving particle agglomeration and floc formation. However, exceeding the optimal dosage can cause particle restabilization and reduced efficiency, as excessive coagulant may lead to charge reversal and diminished removal performance [22]. For neutral coagulants, coagulation-flocculation primarily occurs through polymer bridging, where polymer chains adsorb onto colloidal particles and link them together, independent of particle charges [12]. In non-ionic polymers, this bridging is typically facilitated by hydrogen bonding between the non-ionized sections of the polymer and the functional groups on the surface of the suspended particles.

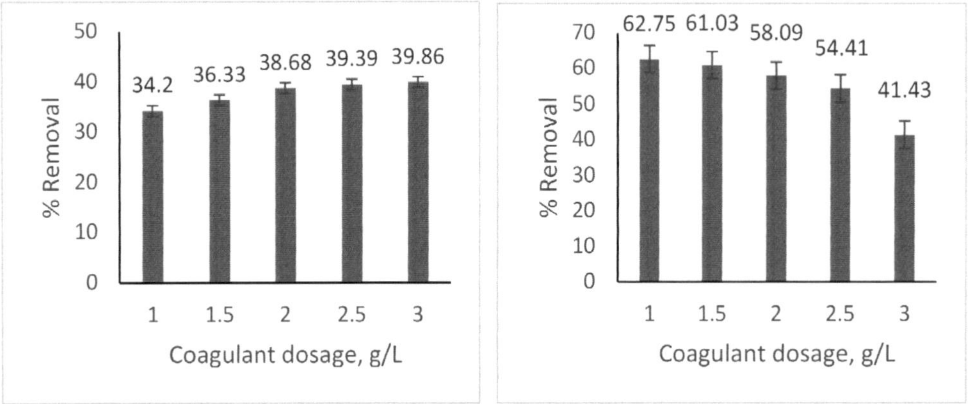

Fig. 3. Effect of coagulant dosage on phosphate removal using unmodified starch (left) and cationic starch (right).

Fig. 3 (right) illustrates that the highest phosphate removal efficiency was achieved with the lowest dosage of cationic starch. The efficiency remained relatively stable within the 1.0 to 1.5 g/L dosage range, with phosphate removal rates of 62.75 % and 61.03 %, respectively. However, phosphate removal decreased from 62.75 % to 41.43 % as the dosage increased from 1.0 to 3.0 g/L at pH 5, likely due to charge reversal and destabilization of colloidal particles from excessive dosing, where excessive cationic species convert negatively charged colloids to positive charges [16]. The results are consistent with [23], which found that optimal phosphorus reduction with cationic starch occurred at doses of 1.0 to 2.0 g/L. At higher doses, performance declined and the solution restabilized. This decrease is due to excessive coagulant causing the formation of positively charged "microflocs" that trap dihydrogen phosphate, which hinders further flocculation and macrofloc formation due to repulsion.

4 Conclusion

This study highlights the effective transformation of pineapple stem waste into a functional biopolymer coagulant for phosphate removal in simulated municipal wastewater. Starch extracted from pineapple stem was successfully modified through etherification with GTAC, resulting in cationic starch with enhanced coagulation performance. FTIR analysis confirmed the presence of quaternary ammonium groups in the modified starch, which contributed to its superior phosphate removal efficiency—achieving up to 63.24 % removal at pH 5 and a low dosage of 1.0 g/L. In contrast, unmodified starch required a higher dosage of 2.5 g/L to reach a maximum removal of 39.39 %. The improved performance of the cationic starch is attributed to its ability to facilitate both charge neutralization and ionic bonding with phosphate ions, unlike the unmodified starch which relied primarily on bridging and adsorption mechanisms. These findings demonstrate that modifying starch from agricultural waste significantly enhances its functionality as a coagulant, offering a sustainable and high-performing alternative to conventional materials. This research provides a strong foundation for further development of bio-based coagulants in water treatment applications, particularly where resource recovery and waste valorization are key priorities.

The authors wish to thank Mapua Institute of Technology at Laguna, Mapua Malayan Colleges Laguna, Chemistry Laboratory, and IFO-CHM Office.

References

1. A. Timofeeva, M. Galyamova, S. Sedykh. Plants **11**, 2119 (2022)
2. Y. Yang, Y. Li, R. Mao, Y. Shi, S. Lin, M. Qiao, X. Zhao. Sep Purif Technol **286**, 120439 (2022)
3. I.W. Almanassra, V. Kochkodan, G. Mckay, M.A. Atieh and T. Al-Ansari. J Environ Manage **287**, 112245 (2021)
4. M. Liu, C. Wang, J. Guo, L. Zhang. J Environ Chem Eng **9**, 106123 (2021)
5. A.A. Owodunni, S. Ismail, S.B. Kurniawan, A. Ahmad, M.F. Imron, S.R.S. Abdullah. J Water Processs Eng **52,** 103573 (2023)
6. S.H. Hong, A.M. Ndingwan, S.C. Yoo, C.G. Lee, S.J. Park. J Environ Manage **270,** 110817 (2020)
7. P. Zhang, M. He, S. Huo, F. Li, K. Li. J Chem Eng **446,** 137081 (2022)
8. R. Mohammadi, M. Hezarjaribi, D.L. Ramasamy, M. Sillanpää, A. Pihlajamäki. J Chem Eng **407**, 126494 (2021)
9. M. Getahun, P. Asaithambi, A. Befekadu, E. Alemayehu. Case Stud Chem Environ Eng **8**, 100370 (2023)
10. S. Ihaddaden, D. Aberkane, A. Boukerroui, D. Robert. J Water Process Eng **49**, 102952 (2022)
11. N. Zaki, A. Charki, N. Hadoudi, O. Fraiha, H. El Ouarghi, A. Salhi, H. Amhamdi, M. H. Ahari. E3S Web of Conf **527**, 03015 (2024)
12. W.L. Ang, A.W. Mohammad. J Clean Prod **262**, 121267 (2020)
13. S. Nimesha, C. Hewawasam, D.J. Jayasanka, Y. Murakami, N. Araki, N. Maharjan. Glob J Environ Sci Manag **81**, 101-116 (2022)
14. A. Nath, A. Mishra, P.P. Pande. Mater Today Proc **46**, 6113-6117 (2021)
15. J. Ren, N. Li, H. Wei, A Li, H. Yang. Water Res **170**, 115361 (2020)
16. S.M. Asharuddin, N. Othman, W.A.H. Altowayti, N.A. Bakar, A. Hassan. Environ Technol Inno **23**, 101637 (2021)
17. A. Villabona-Ortíz, C. Tejada-Tovar, R. Ortega-Toro, N. Licona Dager, M. Millan Anibal. J Water Land Dev **56**, 21-26 (2023)
18. P.H. Chu, M.A. Jenol, L.Y. Phang, M.F. Ibrahim, S. Prasongsuk, W. Bankeeree, H. Punnapaya, P. Lotrakul, S. Abd-Aziz. Chem Biol Technol Agric **8,** 1-15 (2021)
19. N. Nakthong, R. Wongsagonsup, T. Amornsakchai. Ind Crops Prod **105**, 74-82 (2017)
20. A.H. Amran, N.S. Zaidi, K. Muda, L.W. Loan. Int J Eng Sci Technol **7,** 34-37 (2018)
21. H.N. Dayarathne, M.J. Angove, R. Aryal, H. Abuel-Naga, B. Mainali. J Water Process Eng **40**, 101820 (2020)
22. N.A. Zainol, S.N.S. Bakar, S.N. Zailani, A.A. Ghani, K.A. Sohaimi. IOP Conf Ser Mater Sci Eng **932,** 012048 (2020)
23. J. Turunen, A. Karppinen, R. Ihme, Raimo. SN Appl Sci **1**, 1-9 (2019)

E3S Web of Conferences **690**, 03001 (2026) https://doi.org/10.1051/e3sconf/202669003001
ICEPP 2025

Assessing the Impact of Battery Electric Vehicles (BEVs) Adoption on Air Quality and Public Health in Indonesia

Azka Ghaida[1*], *Almira Solehah Ningsih*[1], *Regyna Elrica Dewati*[1], *Revia Muharrami*[1], *Didin Agustian Permadi*[3], *Prachi Goyal*[2], and *Ririn Radiawati Kusuma*[1]

[1]Clean Air Asia, Jakarta, Indonesia
[2]Clean Air Asia, New Delhi, India
[3]National Institute of Technology, Bandung, Indonesia

Abstract. Indonesia faces a major challenge in reducing transport emissions, which contribute significantly to air pollution and increased potential for health risks. Battery Electric Vehicle (BEV) adoption is one of the national strategies to reduce emission from the transportation sector, particularly fine particulate matter ($PM_{2.5}$). This study assesses the potential impacts of BEV adoption on air pollution and health co-benefit in Indonesia. National Vehicle Kilometre Travelled (VKT) data are used as an input for the Long-range Energy Alternatives Planning (LEAP) tool to estimate $PM_{2.5}$ emission and concentration, while the potential health impacts are assessed using AirQ+. Four scenarios with different levels of ambition were analysed: Business as Usual (BaU), scenario I (pessimistic), scenario II (moderate), and scenario III (ambitious), aligned with Indonesia's net-zero target by 2060. Under BaU, $PM_{2.5}$ emissions are projected to increase from 36.8 to 162.9 thousand metric tonnes, with concentrations rising from 20.2 to 95μg/m^3 by 2060. Under BEV adoption scenario, emission and concentration are significantly reduced, potentially preventing up to 110,000 premature deaths under Scenario I and around 700,000 under net zero pathway (Scenarios II and III). These findings highlight the co-benefit of accelerating transport electrification for better air quality and public health in Indonesia.

1 Introduction

Indonesia's road transport has grown significantly over the past two decades, in line with rapid urbanization and increasing daily mobility, particularly in urban areas. Vehicle registration data over the past five years show that motorcycles, passenger cars, and buses experienced comparable growth, with each category increasing by approximately 4%. Motorcycles dominate the national vehicle fleet, accounting for more than 80% of registered vehicles [1].

* Corresponding author: azka.ghaida@cleanairasia.org

© The Authors, published by EDP Sciences. This is an open access article distributed under the terms of the Creative Commons Attribution License 4.0 (https://creativecommons.org/licenses/by/4.0/).

This increase in vehicle numbers has led to significant environmental consequences, particularly increased emissions from fuel combustion. In Indonesia, the transportation sector is one of the largest consumers of fossil energy in Indonesia and contributes significantly to national greenhouse gas emissions. According to the 2022 National Greenhouse Gas Emissions Inventory Report published by the Ministry of Environment and Forestry, emission from the road transport, including motorcycles, passenger cars, and freight vehicles reached approximately 158,938 Gg CO_2e, or about 22% of total energy-sector emissions [2].

Beyond its contribution to climate impacts, transport activity also adversely affects urban air quality. Fossil fuel combustion in vehicles not only emits CO_2 but also air pollutants such as $PM_{2.5}$, SO_x, NO_x, and volatile organic compounds (VOCs), which are key driver of air quality degradation in Indonesia. Emission inventory studies in several Indonesian cities indicate that the transportation sector as one of major sources of $PM_{2.5}$ emissions [3, 4]. Exposure to this pollutant is strongly associated with adverse health outcomes, including increased risks of cardiovascular disease, respiratory disorders, and premature mortality [5].

These interlinkages between climate change and air quality highlight the co-benefits of mitigation strategies, whereby efforts to reduce CO_2 emissions through reduce fossil fuel consumption also led to air pollutant reduction. Building on these co-benefits, Indonesia has committed under the Paris Agreement to achieve Net Zero Emissions (NZE) by 2060 through its Enhanced Nationally Determined Contribution (ENDC), with transport decarbonization identified as a key sector. This commitment is further reinforced by Presidential Regulation No. 55 of 2019 [6] on the Acceleration of Battery Electric Vehicle (BEV) Programs for road transport.

Despite growing policy commitment, empirical evidence on quantifying the impacts of BEV adoption on air quality and public health in Indonesia remains limited. However, evidence from other countries have demonstrated that increased BEV adoption can significantly reduce $PM_{2.5}$ concentration [7] and substantially mitigate associated health impacts. Therefore, this study is designed to address this evidence gap by providing empirical insights to support the formulation of policies aimed at accelerating vehicle electrification.

2 Methodology

2.1 Study Design and Scopes

This study is conducted at the national level using a scenario-based model design to evaluate the potential impact of BEV adoption on air pollution and public health. Baseline data from 2023 is used, based on Statistics Indonesia (BPS), with the projection extended to 2060. The study framework integrated three components: (i) on-road transport data activity, (ii) emission and air pollutant concentration estimated using LEAP-IBC, and (iii) health outcomes quantified with AirQ+.

Four scenarios (BaU and three (3) intervention scenarios) were considered based on Indonesia's national policy target and existing transport studies (Table 1). All scenarios were projected until 2060, aligning with the national net zero emission target, including several interim checkpoints that reflect progressive policy targets and technology adoption pathways. Three intervention scenarios were developed with different level of ambitious. Scenario I, representing the pessimistic case, refers to the 2025 national target set by the Ministry of Energy and Mineral Resources, with less than 35% BEV adoption across all vehicle types by 2060. Scenario II represents the moderate pathway, targeting net-zero emissions by 2060, while Scenario III is the most ambitious, aiming to achieve net-zero by 2040. The BEV target refers to the projected percentage of BEVs used compared to overall ICE vehicles. These checkpoints are estimated based on the government's sectoral planning pathway.

Table 1. Scenario Interventions Developed Based on Existing Policies.

Scenario Intervention	BEV Target		
	Motorcycles	Private Cars	Bus
Business as Usual (BAU)	Based on vehicle growth (with no intervention)		
Scenario I (pessimistic)	2030: 2%; 2060: 29%	2030: 2%; 2060: 32%	2030: 2%; 2060: 23%
Scenario II (moderate)	2030: 11%; 2060: 100%	2030: 7%; 2060: 98%	2030: 4%; 2060: 95%
Scenario III (ambitious)	2030: 7%; 2040: 100%	2030: 8%; 2040: 100%	2060: 100%

2.2 Transport Data Activity

Transport activity is calculated based on vehicles-kilometres travelled (VKT) value, estimated from the number of registered vehicles and the daily average VKT per province. The VKT is disaggregated by vehicle type (motorcycles, passenger cars, and buses) and by regional classification (urban, suburban), with the projection of vehicle growth derived using historical data. Due to data limitations, VKT is estimated using values from a previous study in Jakarta as representative of the urban area and Aceh Province as representative of the suburban area, to obtain a national approximation [8, 9]

2.3 Emissions and Ambient $PM_{2.5}$ Estimation

Emission of $PM_{2.5}$ and other relevant precursors is calculated using the LEAP-IBC tool, a national-scale integrated modelling framework that tracks energy consumption and emission and evaluates potential health co-benefit [10]. Emission factors were derived from EMEP/EEA and IPCC guidelines. In this study, annual emissions from the transport sector were projected for each scenario for comparison across the electrification pathways.

LEAP-IBC is also used to estimate the national average ambient $PM_{2.5}$ concentration, which is calculated from the estimated $PM_{2.5}$ emission load. The model calculates population-weighted $PM_{2.5}$ concentration at the national level, providing an aggregate indicator of exposure. As the data input is limited to the transport sector, these concentrations should be interpreted as a predictive value representing only air quality conditions from this sector, not all emission sources. The output estimates were then used as input for the health impact assessment.

2.4 Health Impact Assessment

Health impacts were quantified using WHO's AirQ+ model, with $PM_{2.5}$ concentration outputs from LEAP-IBC used as an input. As this study is only using transportation sector data, using LEAP-IBC alone could be less accurate on health impact estimates, as the model is designed to incorporate data from all sectors. In addition to ambient data concentration, AirQ+ requires baseline and target year of analysis, total population, total population affected, and coverage area [11]. The model output includes a proportion of air pollution-related health impact rates, estimated number of attributable cases of premature deaths, and estimated attributable proportion per 100,000 people within the population at risk.

2.5 Uncertainty Consideration

This study focuses on passenger vehicles, including motorcycles, passenger cars, and buses, and does not include freight transport. The analysis is limited to vehicle electrification scenarios and does not consider other transport-related policies, such as public transport improvements or vehicle age restrictions.

Emission estimates are based on fuel combustion emissions from the transport sector using the LEAP-IBC model, without accounting for fugitive emissions or emissions from other sectors. Vehicle growth and BEV adoption projections are derived from historical trends, and uncertainties associated with these assumptions were not quantified. In addition, national energy transition policies and power sector dynamics are not incorporated into the analysis.

3 Result & Discussion

3.1 Baseline Emission and Concentration

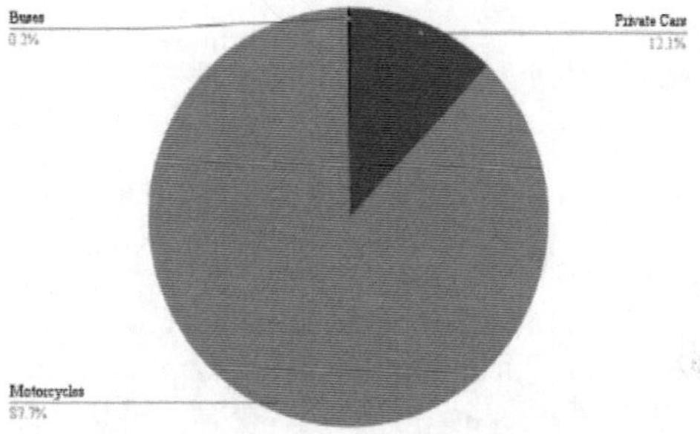

Fig. 1. Proportion Number of Vehicles 2023

In 2023, the number of vehicles in Indonesia is largely dominated by motorcycles, followed by private cars and buses (Figure 1). The small share of buses is due to the limited fleet size compared to private vehicles, as public transport services in many cities are still underdeveloped [1]. The Business as Usual (BaU) scenario assumes no intervention to shift from ICE to BEV. Under this scenario, the ICE vehicle is projected rise constantly between 2023 and 2060 by around 343% driven by rapid motorization and limited vehicle technology improvement, including fuel quality [12]. This growth also not consider future policies, such as increasing mode shift to public transport or stricter emission regulation, which may reduce emissions. Those policies are excluded in this model due to potential infrastructure gaps and uneven implementation capacity across regions in Indonesia [13].

Under the BaU scenario, transport-related emissions in 2060 are projected to be nearly four times higher than in the 2023 baseline (Figure 2a), primarily driven by motorcycles for about 50% of the total emissions, followed by private cars and buses (Figure 2b). This increase leads to a severe decline in air quality, especially $PM_{2.5}$ concentrations caused by transportation activity, with the estimated annual average of $PM_{2.5}$ in 2060 increasing to 85 mg/m3 in 2060 (Figure 2c), far exceeding Indonesia's recommended annual interim target of 15 mg/m3. The high concentration results on LEAP-IBC is not only due to primary $PM_{2.5}$ concentration, but also to secondary formation of other pollutants. LEAP also calculated emissions from electrical generation through well-to-wheel approach, resulting higher estimated $PM_{2.5}$ concentration. The models do not include the GoI's planned on energy

https://doi.org/10.1051/e3sconf/202669003001

transition to cleaner energy sources, as the country still relies heavily on coal for electricity generation.

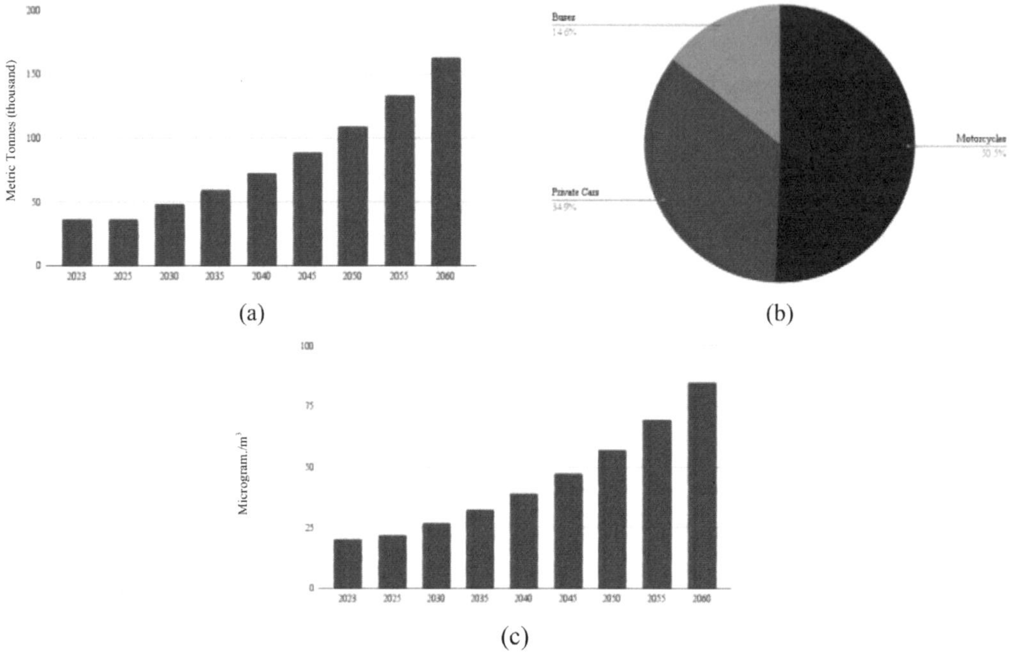

Fig 2. (a) $PM_{2.5}$ emission under baseline scenario (BaU), (b) $PM_{2.5}$ emission proportion per vehicle type, (c) Projection of $PM_{2.5}$ concentration under baseline scenario.

3.2 Effect of BEV Adoption Scenario

Compared to the BaU scenario, where emissions continue to rise along with vehicle growth, the adoption of BEVs significantly reduces $PM_{2.5}$ emissions across all intervention pathways (Figure 3a). In scenario I, emissions reduction is relatively small, continues to increase until 2040, before slowly stabilizing at 100,000 MT by 2060. In scenario II, emissions begin to decline after 2035 as an initial target and reach near-zero by 2060. Meanwhile, for Scenario III, as the most ambitious intervention target, the projection shows a steeper decrease around 2030 and approaches zero emissions well before 2050.

In both scenarios II and III, the transition to BEV significantly reduces in transport-related $PM_{2.5}$ emissions. BEV adoption is considered to produce zero $PM_{2.5}$ emission from the tailpipe [14], while gasoline and diesel vehicles cause an average of 1.57 mg/km and 57.8 mg/km [15]. A slight increase of $PM_{2.5}$ emissions from 2030 to 2035 shows under scenario III, indicates vehicle growth similar to the baseline trend. However, the sharp decline after 2035 marks a strong reduction, driven by an ambitious scenario that targets 100% electrification on motorcycles and passenger cars by 2040. As a result, the transport sector achieved the net-zero emission target before 2060.

https://doi.org/10.1051/e3sconf/202669003001

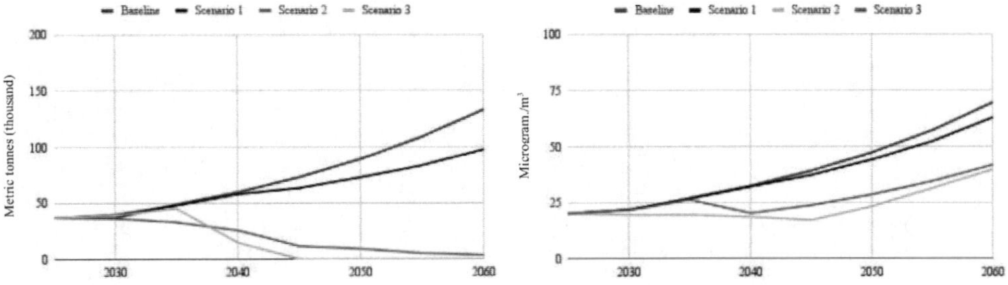

(a) (b)

Fig 3. (a) Environmental Effects ($PM_{2.5}$ Emissions) in Physical Units, (b) Projection of $PM_{2.5}$ concentration under BEV scenarios.

To assess the impact, $PM_{2.5}$ concentration is needed to estimate pollutant exposure, rather than total emissions. The concentration data, combined with population data, were then analyzed to determine how BEV adoption influences population-weighted $PM_{2.5}$ levels. The result shows that all scenario interventions significantly reduce of $PM_{2.5}$ concentration relative to the BaU (Figure 3b). Without interventions, the reliance on fossil fuel-based transportation will lead to a persistent increase in $PM_{2.5}$ concentrations [16]. Among the scenarios, scenario I shows limited improvement of $PM_{2.5}$ concentration, while the other two scenarios demonstrate significant reduction.

In scenario II, $PM_{2.5}$ concentrations gradually decline from the base year to their lowest level around 2045, before increasing again toward 2060, but remaining below the projected under BaU. Scenario III shows a more significant decline in concentration through 2040, followed by a similar upward trend in 2060. This pattern reflects a trade-off—as full electrification progresses, energy consumption shifts to the power sector. While the concentration levels remain far below the BaU, the benefits would be greater if the transition to cleaner power plants is accelerated. These findings align with the previous study highlighting that, without decarbonizing the electricity supply, air quality improvements from EV adoption are hindered by power sector emissions [17].

3.3 Health Benefits of BEV Adoption

In line with the projected decline in $PM_{2.5}$ concentrations, AirQ+ results show that all intervention scenarios result in significant reductions in health impact, especially after 2040. In contrast, projections under the BaU scenario show a steady increase in health burden over time. This study used three main health impact parameters: (i) the proportion of health impacts due to air pollution, as the total health burden caused by air pollution (Figure 4a); (ii) the number of premature deaths due to air pollution (≥ 30 years), representing the absolute number of premature deaths in the 30+ age group (Figure 4b); and (iii) the proportion attributable per 100,000 at-risk population (≥ 30 years), which indicates the premature death rate attributable to air pollution per 100,000 individuals in that age group.

Under the BaU, attributable proportion from air pollution rises from 11% in 2023 to 46% by 2060. Scenario I shows slightly lower but still substantial growth, reaching 42% in 2060. Meanwhile, scenarios I and II demonstrate significant differences, with proportions of 29% and 30% in 2060, respectively (Figure 2a).

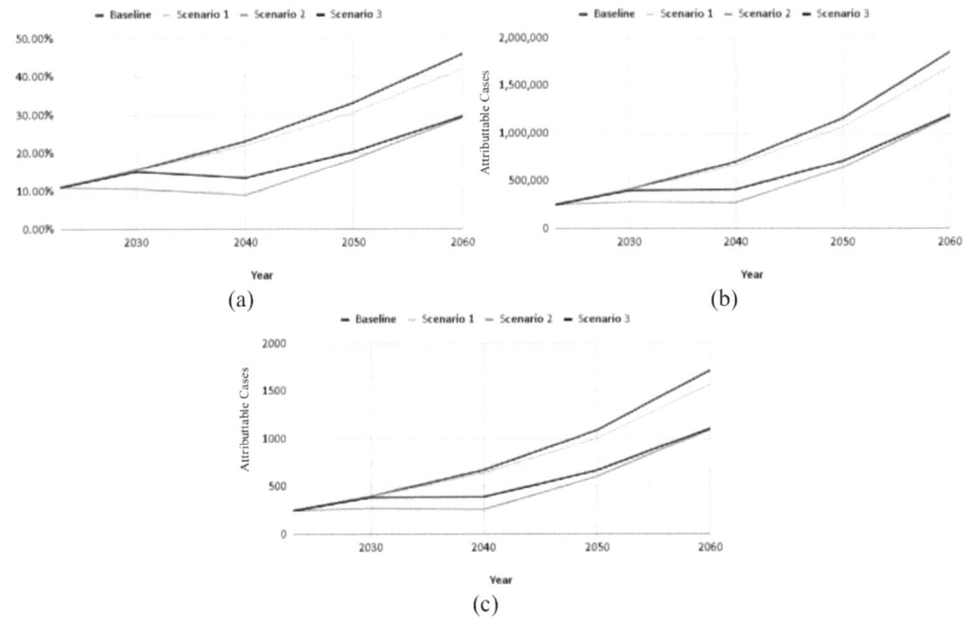

Fig. 4 (a) The Attributable Proportion of Air Pollution-related Health Impacts, (b) Estimated number of Attributable Cases of Premature Mortality (≥ 30 years), (c) Estimated Attributable Proportion per 100,000 Population at Risk (≥ 30 years).

By 2060, premature deaths are projected to reach 1.8 million under BaU scenarios and 1.69 million under scenario I. In contrast, scenarios II and III reduce to around 1.18 and 1.19 million premature deaths, which means that nearly 700,000 premature deaths could be avoided (Figure 2b). A similar pattern is observed in the attributable proportion per 100,000 population, while the BaU scenario projects 1,716 cases, scenario I reduces this to 1,570 cases, and scenarios II and III show larger reductions to 1,096 and 1,108 cases by 2060 (Figure 2c).

These results show that all three scenarios' interventions have significantly reduce $PM_{2.5}$ emissions, leading to a lower health burden and mortality risk associated with air pollution exposure. Cross-scenario comparisons further demonstrate that more ambitious policies promoting the adoption of battery electric vehicles (BEVs) result in substantially greater public health benefits [18].

4 Conclusion

This study demonstrates that the adoption of BEV can significantly improve air quality and deliver health co-benefits in Indonesia. Across all intervention scenarios, reduction in both $PM_{2.5}$ emissions and concentration are observed, with scenarios II and III showing significant results for achieving net-zero emissions target by 2060. Under most ambitious pathway, Scenario III, $PM_{2.5}$ concentration are projected to decline by approximately 60% by 2060 compared to BaU scenario, potentially preventing nearly 700,000 premature deaths linked to long-term PM2.5 exposure by 2060. However, the finding also indicates that transport electrification must be accompanied by an accelerated energy transition toward cleaner electricity generation to maximize its positive impacts.

Future research should apply an integrated analytical framework that incorporates national energy transition policies and power sector dynamics, enabling more robust assessment to support data-driven policy formulation.

This research is conducted with the support of ViriyaENB, through the Clean Air Asia Indonesia project. The findings and interpretation are those of the author and should not represent the donor.

References

1. Badan Pusat Statistik, *"Jumlah Kendaraan Bermotor Menurut Provinsi dan Jenis Kendaraan (unit)*, [2023]," 2023. [Online]. Available: https://www.bps.go.id/id/statistics-table/3/VjJ3NGRGa3dkRk5MTlU1bVNFOTVVbmQyVURSTVFUMDkjMw==/jumlah-kendaraan-bermotor-menurut-provinsi-dan-jenis-kendaraan--unit---2023.html. [Accessed May 2025].
2. Direktorat Jenderal Pengendalian Perubahan Iklim, Direktorat Inventarisasi Gas Rumah Kaca dan MPV, *"Laporan Inventarisasi Gas Rumah Kaca (GRK) dan Monitoring, Pelaporan, Verifikasi (MPV),"* Kementerian Lingkungan Hidup dan Kehutanan, [2024].
3. P. Lestari, *"Study Emission Inventory in Jabodetabek. Jakarta: Institut Teknologi Bandung.,"* [2025].
4. M. Santoso, D. Lestiani, S. Kurniawati, E. Damastuti, I. Kumartini, D. Atmodjo and D. K. Sari, *"Assessment of urban air quality in Indonesia,"* Aerosol and Air Quality Research, vol. 20, no. 10, [2020].
5. World Health Organization, *"WHO global air quality guidelines. Particulate matter (PM2.5 and PM10), ozone, nitrogen dioxide, sulfur dioxide and carbon monoxide,"* World Health Organization, [2021].
6. Pemerintah Pusat Indonesia, *"Peraturan Presiden (Perpres) Nomor 55 Tahun 2019 tentang Percepatan Program Kendaraan Bermotor Listrik Berbasis Baterai (Battery Electric Vehicle) untuk Transportasi Jalan,"* Pemerintah Pusat Indonesia, [2019].
7. Y. Wu, F. Gu, Y. Ji, S. Ma and J. Guo, *"Electric vehicle adoption and local PM2. 5 reduction: Evidence from China.,"* Journal of Cleaner Production, vol. 396, no. 136508, [2023].
8. M. Gidarjati and T. Matsumoto, *"Dynamic Vehicle Age-Based Cohort Model to Estimate the Emission from the Transportation Sector in Jakarta,"* Ecological Engineering & Environmental Technology, vol. 25, no. 4, [2024].
9. R. Rahma, M. Mariana, A. Gani, S. Sugiarto and M. Isya, *"Assessing carbon monoxide (CO) emissions from private vehicles: A pathway to cleaner urban mobility,"* IOP Conf. Series: Earth and Environmental Science, vol. 1479, [2025].
10. J. C. I. Kuylenstierna, C. Heaps, C. Malley, H. Vallack and K. Hicks, *"The Long-range Energy Alternatives Planning – Integrated Benefits Calculator (LEAP-IBC)"*, Stockholm Environment Institute, [2017].
11. World Health Organization, *"Health impact assessment of air pollution: introductory manual to AirQ+"*, World Health Organization, Regional Office for Europe, [2020].
12. G. Wang, W. Xia, Y. Xiao, X. Guan and X. Zhang, *"The Synergy Between CO2 and Air Pollution Emissions in Chinese Cities by 2060: An Assessment Based on the Emissions Inventory and Dynamic Projection Model," Sustainability,* vol. 16, no. 9338, [2024].

13. P. Sapbamrer, P. Assavanopakun and J. Panumasvivat, *"Decadal trends in ambient air pollutants and their association with COPD and lung cancer in upper Northern Thailand: 2013–2022,"* Toxics, vol. 12, no. 5, [2024].
14. H. Togun, A. Basem, N. Biswas, T. Abdulrazzaq, A. M. Abed, J. M. Dhabab, A. Chattopadhyay,, K. Slimi, D. Paul, P. Barmavatu and A. Chrouda, *"Development and comparative analysis between battery electric vehicles (BEV) and fuel cell electric vehicles (FCEV),"* Applied Energy, Elsevier, vol. 388(C), [2025].
15. H. H. Yang, N. B. Dhital, L. C. Wang, Y. S. Hsieh, K. T. Lee, Y. T. Hsu and S. C. Huang, *"Chemical Characterization of Fine Particulate Matter in Gasoline and Diesel Vehicle Exhaust,"* Aerosol and Air Quality Research, vol. 19, no. 1439–1449, [2019].
16. Y. Xiong, D. Partha, N. Prime, S. J. Smith, N. Mariscal, H. Salah and Y. Huang, *"Long-term trends of impacts of global gasoline and diesel emissions on ambient PM2.5 and O3 pollution and the related health burden for 2000–2015," Environmental Research Letters,* vol. 17, no. 10, [2022].
17. N. Damanik, R. Saraswani, F. D. Hakam, and Dea M and D. M. Mentari, *"A Comprehensive Analysis of the Economic Implications, Challenges, and Opportunities of Electric Vehicle Adoption in Indonesia,"* Energies, vol. 18, no. 1384, [2025].
18. E. Y. Lim and G. D. Kim, *"Particulate matter-induced emerging health effects associated with oxidative stress and inflammation.,"* Antioxidants *13,* vol. 13, no. 10, [2024].

Investigation of heavy metals in indoor dusts of households in an indubitable range from a landfill located in south-western India

Naveen S. Lal[1], *C.T* Aravindakumar[2,3], and *Usha K.* Aravind[1*]

[1]School of Environmental Studies, Cochin University of Science and Technology, Kochi, India
[2]School of Environmental Sciences, Mahatma Gandhi University, Kottayam, India
[3]Inter University Instrumentation Centre, Mahatma Gandhi University, Kottayam, India

Abstract. Waste treatment plants, landfills, and dump yards are surplus in trash that emits various elements, including heavy metals, into the atmosphere either by human activities or by any fire events. This study investigates the heavy metals in indoor dusts of households situated at various directions and varying distances away from the landfill and its impact on human health and the environment. The concentration of heavy metals in dust samples followed the order (mg/kg): Fe (14905.68) > Mn (377.57) > Cu (275) > Zn (219.41) > Cr (145.91) > Pb (124.86) > Ni (68.89) > Cd (5.58). The contamination factor underscores elevated pollution for Zn and Pb (marker elements) and pollution load index showed contamination in most of the sites. While considering health risk assessments, the Hazard Index (HI) for non-carcinogenic effects on both children and adults indicates values < 1. Carcinogenic effects possessed values in the appreciable category except for dermal exposure for Cd, where the values fell in the acceptable risk category, suggesting no immediate harm to the exposed population. The study postulates that the high pollution index levels may hint at potential future risk. The study also demonstrates the significance of risk assessment formulas for immediate health risk assessment.

1 Introduction

Metropolitan expansion, industrial development, and demographic boom have led to a rise in garbage production worldwide, especially in developing nations, where improper waste management has become a significant concern for both humans and the environment. As we all know, the increasing population in urban areas leads to an increase in consumption, which in turn leads to an increase in the generation of waste materials, especially solid waste that is difficult to dispose of. The most prominent and easy method of solid waste disposal is the use of sanitary landfills. Most of the solid wastes generated (E-waste, plastic waste, food waste and all other paper wastes) in urban areas are dumped at landfills, and this landfill acts as a reservoir of harmful pollutants. These pollutants can be released from landfills to the atmosphere due to various activities in landfills such as transit, buildup, squeezing and

* Corresponding author: uka@cusat.ac.in

© The Authors, published by EDP Sciences. This is an open access article distributed under the terms of the Creative Commons Attribution License 4.0 (https://creativecommons.org/licenses/by/4.0/).

annihilation. Unexpected fires which occur in landfills by internal (heat generation from biological degradation processes, methane generation from anaerobic decomposition of organic waste, spontaneous ignition) and external sources (excessive temperature during summer, sparks from faulty landfill machinery, etc) may also result in emission of pollutants.

The pollutants emitted from the landfills include numerous organic (Dioxins, furans, polycyclic aromatic hydrocarbons, perfluorinated alkyl substances, polymers, BTEX and other hydrocarbons) and inorganic pollutants (crustal elements, heavy metals and gases such as CO, CO_2, SO_X, NO_X). Landfills are abundant sources of heavy metals, and their emission to the atmosphere may lead to exposure to the human population residing nearby, which in turn causes serious health effects to the exposed population. Even though landfill aerosol emissions and their characterisation have been formerly studied, studies on aerosol characterisation in indoor microenvironments surrounding the landfill or at certain distances away from landfills are lacking.

So, in order to address these issues, this study focusses mainly on (a) target analysis of heavy metals present in indoor dust of households nearby and certain range away from a waste treatment plant (treatment plant often remain as landfills or dump yards), (b) assessing the contamination level of the studied households with help of various pollution indices, (c) assessing non-carcinogenic and carcinogenic human health risk of the population residing in the sampled households.

2 Materials and Methods

2.1 Study area and sample collection

The study area consists of a municipal solid waste treatment plant (Brahmapuram Solid Waste Treatment Plant (BSWTP)) located at Brahmapuram (also known as Brahmapuram landfill) and 17 other places encircling the plant altogether, located in the Ernakulam district of Kerala state, India. The study area is shown in Figure 1. Kochi city, known as the financial and economic capital of Kerala, located in the Ernakulam district, generates a huge quantity of waste per year that ends up in the BSWTP.

A total of 17 representative dust samples were taken from households of various directions (north, northeast, northwest, south, southeast, east, west, and southwest) and with varying distances (0-10 km, 10-20 km, and 20-30 km land distance) from BSWTP. Sampling was carried out between 4 and 14 March, 2023, during a catastrophic fire event at BSWTP. The weather was hot, and no rainfall was observed during the sample collection period. The samples were collected by sweeping the objects containing dust with a plastic-type brush from each household. The collected dust sample was further covered in aluminium foil and was immediately transported to the laboratory and stored at 4°C. After sampling from each site, the sampling brush was washed with acetone to prevent contamination.

2.2 Sample preparation and analysis

The samples were air dried in a hot air oven at a temperature of 30ºC for 2-3 hours and were sieved using a sieve having a mesh size of 63 μm. About 0.1 g of each sieved dust was weighed and transferred to a 250 ml beaker. 10 ml of ultrapure nitric acid (Signa-Aldrich HNO3 70% purified by redistillation) was added, followed by the addition of ultrapure water till it reached the 100 ml mark. Hot plate digestion with a temperature of 95ºC was performed on the prepared solution till the solution reached the 50 – 60 ml mark. Further, 2-3 drops of 30% H2O2 were added. The beaker containing the solution was covered with aluminium foil (holes were made in the foil for vapour escape) and was kept aside to attain room temperature.

This cooled solution was filtered twice, beginning with Whatman filter paper No. 1 (pore size 11 μm), followed by membrane filtration (pore size 0.2 μm). The filtered solution was stored in a small Teflon bottle until it was subjected to ICP-MS analysis (Thermoscientific iCAP-Q ICP-MS).

2.3 Pollution Indices

In order to determine the level of contamination, various pollution indices were introduced. Two such types of pollution indices, such as Contamination Factor (CF) and Geo-accumulation index (Igeo), were computed in this study to determine the level of pollution in the collected household dusts.

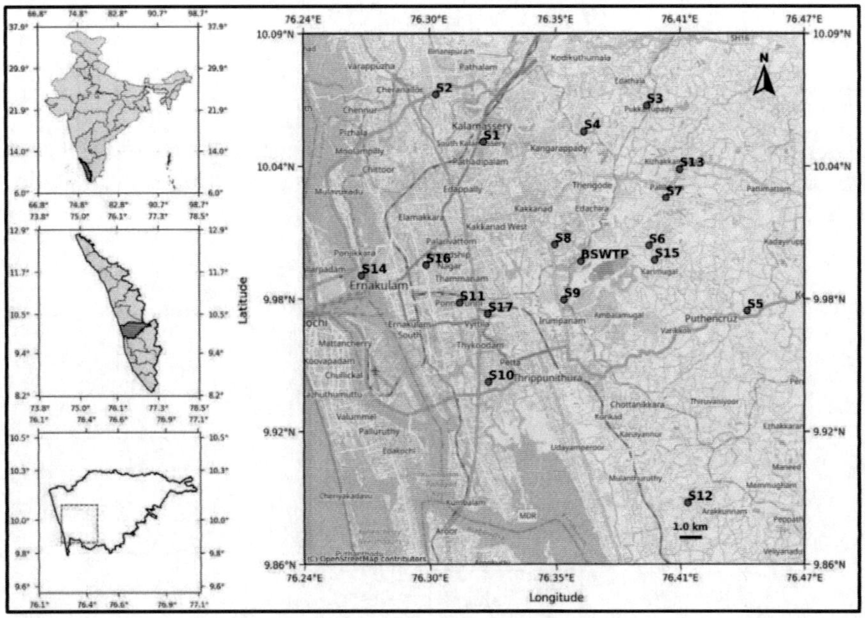

Fig. 1. Study area depicting the BSWTP and sampling points surrounding BSWTP

2.3.1 Contamination factor (CF)

The contamination factor was computed for dust samples from various household using equation (1) [1].

$$CF = C_n/B_n \qquad (1)$$

Where CF = Contamination factor

C_n = concentration of heavy metal 'n' in a particular indoor microenvironment

B_n= Indian natural soil background concentration of heavy metal 'n' (except Cd)

B_n values for various heavy metals are 32015 μg/g for Fe, 56.5 μg/g for Cu, 209 μg/g for Mn, 22.1 μg/g for Zn, 27.7 μg/g for Ni, 13.1 μg/g for Pb, 114 μg/g for Cr .

For Cd, 6 μg/g was taken as the background concentration due to unavailability of data (maximum limit for Cd in Indian soils)

The classification of contamination factors according to pollution is shown in Table 1

Table 1. Contamination factor classification [1]

Range	Level of contamination
CF < 1	Low contamination
1< CF< 3	Moderate contamination
3< CF < 6	Considerable contamination
CF > 6	Very high contamination

2.3.2 Pollution Load Index (PLI)

Pollution load index was used for finding the overall contamination caused by heavy metals in the individual households. The equation (2) was utilised for finding the pollution index. The description of PLI values are shown in Table 2.

$$PLI = (CF_1 * CF_2 * CF_3 * \ldots\ldots * CF_n)^{1/n} \quad (2)$$

Where,

CF_1 = Contamination factor of site 1
CF_2= Contamination factor of site 2
CF_3 = Contamination factor of site 3
CF_n = Contamination factor of site n

Table 2. Pollution load index [1].

PLI values	Description
PLI < 1	No overall pollution
PLI > 1	Site contamination

2.4 Human health risk assessment

Human health risk assessment was computed based on non-carcinogenic risk and carcinogenic risk [2]. In terms of non-carcinogenic risk, exposure to the studied heavy metals occurs through three main pathways, namely inhalation, ingestion, and dermal exposure. Average Daily Dose (ADD) for each exposure pathway was computed, followed by computation of Hazard Quotient (HQ) for the three exposure pathways. The Hazard Index (HI) was finally derived from the calculated HQs from the three exposure pathways. Separate values were assigned for exposure assessment for children and adults in terms of ADDs, HQs, and HIs.The ADD for inhalation, ingestion, and dermal exposure was calculated from the following equations (3), (4), and (5).

$$ADD_{inh} = (C * InhR * EF * ED)/(PEF * BW * AT) \quad (3)$$

$$ADD_{ing} = (C * IngR * CoF * EF * ED)/ (BW * AT) \quad (4)$$

$$ADD_{dermal} = (C * SA * CoF * AF * ABF * EF * ED)/(BW * AT) \quad (5)$$

Where,
ADD_{inh} = Average daily dose for inhalation exposure
ADD_{ing} = Average daily dose for ingestion exposure
ADD_{der} = Average daily dose for dermal exposure
The values for the variables in equation (3), (4) and (5) are shown in Table 3

Table 3. Parameters regarding non-carcinogenic and carcinogenic human health risk assessment [2].

Factor	Definition	Unit	Value (Adult)	Value (Children)
C	Concentration of heavy metals in indoor dusts	mg/kg	-	
InhR	Inhalation rate	m^3/kg	12.8	7.63
PEF	Particle emission factor	m^3/kg	$1.36 * 10^9$	$1.36 * 10^9$
IngR	Ingestion rate	mg/day	200	100
EF	Exposure frequency	days/year	365	365
ED	Exposure duration	years	24	6
BW	Body weight	kg	55.9	15
AT	Average time	days	365 * ED	365 * ED
CoF	Conversion factor	kg/mg	$1 * 10^{-6}$	$1 * 10^{-6}$
ABF	Dermal absorption factor	-	0.001	0.001
SA	Surface area of skin that contacts dust	cm^2	4350	1600
AF	Skin adherence factor	mg/cm^2	0.7	0.2

Hazard Quotient for inhalation, ingestion, and dermal exposure was calculated from the equations (6), (7), and (8). If HQ ≤ 1, there are no harmful health effects, and if HQ > 1, there are certain health effects [2].

$$HQ_{inh} = ADD_{inh}/(RfD_{inh\text{-}n}) \quad (6)$$
$$HQ_{ing} = ADD_{ing}/(RfD_{ing\text{-}n}) \quad (7)$$
$$HQ_{der} = ADD_{der}/(RfD_{der\text{-}n}) \quad (8)$$

$RfD_{inh\text{-}n/ing\text{-}n/der\text{-}n}$ = Reference dose of the designated heavy metal 'n'. RfD is an acceptable limit for human hazard assigned for certain metals during regular exposure. RfD values assigned for inhalation, ingestion, and dermal exposure of various heavy metals are shown in the Table 4.

Hazard Index was calculated from the equation (9)

$$HI = \Sigma HQ_i \quad (9)$$

i = route of exposure (inhalation, ingestion and dermal exposure)

If HI < 1, there is no non-carcinogenic risk for humans, and if HI > 1, there is a certain human non-carcinogenic risk.

For carcinogenic risk assessment, Lifetime Average Daily Dose (LADD) for inhalation, ingestion, and dermal exposure was computed, followed by computation of Carcinogenic Risk (CR) for the three routes of exposure. From CR, Total Carcinogenic Risk (TCR) was finally estimated [2]. Out of the metals estimated in this study, only Cu, Cd, Pb, Ni, and Cr are considered carcinogenic elements, and out of these elements, the computation of CR of Cd, Cr, Pb, and Ni was only estimated due to the unavailability of the slope factor for other elements. Out of Cd, Cr, Pb, and Ni, the element Cd was the only one subjected to CR of all routes of exposure due to the availability of data in all three exposure routes, and the rest of the elements were subjected to CR_{inh} only.

Table 4. Reference dose values for the studied heavy metals

Element	**RfD_{inh}**	**RfD_{ing}**	**RfD_{der}**
Pb	$3.52 * 10^{-3}$	$3.5 * 10^{-3}$	$5.25 * 10^{-4}$
Cd	$1* 10^{-3}$	$1 * 10^{-3}$	$1* 10^{-5}$
Cr	$2.86 * 10^{-5}$	$3 * 10^{-3}$	$6 * 10^{-5}$
Mn	$1.43 * 10^{-5}$	$4.60 * 10^{-2}$	$1.84 * 10^{-3}$
Ni	$2.06 * 10^{-2}$	$2* 10^{-2}$	$5.4 * 10^{-3}$
Fe	$8 * 10^{-1}$	$7 * 10^{-1}$	$7 * 10^{-1}$
Zn	$3 * 10^{-1}$	$3 * 10^{-1}$	$6 * 10^{-2}$
Cu	$4.02 * 10^{-2}$	$4 * 10^{-2}$	$1.20 * 10^{-2}$

LADD for three routes of exposure was calculated from the equations (10), (11) and (12)

$$LADD_{inh} = (((C * EF) / (AT * PEF)) * (((InhR_{child} * ED_{child}) / BW_{child}) + ((Inh_{adult} * ED_{adult}) / BW_{adult}))) \quad (10)$$

$$LADD_{ing} = (((C * EF) / AT) * (((IngR_{child} * ED_{child}) / BW_{child}) + ((IngR_{adult} * ED_{adult}) / BW_{adult})) * 10^{-6}) \quad (11)$$

$$LADD_{der} = (((C * EF * ABF) / AT) * (((AF_{child} * SA_{child} * ED_{child}) / BW_{child}) + ((AF_{adult} * SA_{adult} * ED_{adult}) / BW_{adult})) * 10^{-6}) \quad (12)$$

From LADDs, CR for three routes of exposure was calculated from the equations (13), (14) and (15)

$$CR_{inh} = LADD_{inh} * SF_{inh} \quad (13)$$

$$CR_{ing} = LADD_{ing} * SF_{ing} \quad (14)$$

$$CR_{der} = LADD_{der} * SF_{der} \quad (15)$$

Where $LADD_{inh}$, $LADD_{ing}$, and $LADD_{der}$ stand for lifetime average daily doses for inhalation, ingestion, and dermal exposure, while CR_{inh}, CR_{ing}, and CR_{der} stand for carcinogenic risks for inhalation, ingestion, and dermal exposure. SF_{inh}. SF_{ing}, SF_{der} implies slope factor for inhalation, ingestion, and dermal exposure, which varies with the element being calculated. The values for SF regarding various routes of exposure for various elements are provided in Table 5.

From the equations (13), (14) and (15), TCR was calculated (for Cd only) as shown in equation (16)

$$TCR = \Sigma\, CR_i \qquad (16)$$

Table 5. Slope factor values for various elements in various modes of exposure [2]

Slope factor	Cr	Cd	Ni	Pb
SF_{inh}	42	6.30	0.84	0.0085
SF_{ing}	-	6.10	-	-
SF_{der}	-	6.10	-	-

Where i = route of exposure (inhalation, ingestion, and dermal exposure).

It is assumed that if CR/TCR < $1*10^{-6}$, it is regarded as 'appreciable,' and if CR/TCR values fall between $1*10^{-6}$ and $1*10^{-4}$, it is regarded as 'acceptable or tolerable risk.' If CR/TCR values exceed $1*10^{-4}$, then it is assumed that 1 in 10,000 people may develop any type of cancer from lifetime exposure to carcinogenic hazards [2].

3 Results and Discussions

3.1 Heavy metal concentration in indoor dust

The results of the analysed dust revealed that the total concentration of heavy metals (mg/kg) from various sites followed the pattern:S4 (5214.29) > S8 (3254.44) > S9 (2875.30) > S13 (2761.90) > S6 (2729.71) > S5 (2460.98) > S2 (2274.37) > S3 (2273.01) > S7 (1984.76) > S16 (1834.37) > S1 (1583) > S11 (1486.89) > S15 (1467.93) > S12 (866.47) > S14 (446.33) > S10 (418.42) > S17 (328.99), and in terms of concentration of heavy metals (mg/kg) from the entire sites, it followed the pattern:Fe (14905.68) > Mn (377.57) > Cu (275) > Zn (219.41) > Cr (145.91) > Pb (124.86) > Ni (68.89) > Cd (5.58). This pattern of heavy metal concentration was almost the same as the pattern of heavy metals observed in household dusts in urban sites from a previous study, which had Zn, Pb, Cr, and Cu with high concentrations [2].

The present study identified Fe as the heavy metal with the highest concentration and Cd the lowest. Numerous studies conducted worldwide previously have found cadmium as the lowest among all the heavy metals studied. But Fe has not been frequently studied in indoor dust; therefore, comparison of the obtained concentration with previous studies was not possible. A study carried out on indoor air (PM_{10} and $PM_{2.5}$) of urban and rural households of Agra, India, reported Fe concentration, in which Fe was the highest observed concentration in households situated in both urban and rural sites. Crustal sources and burning of scrap and mixed wastes are identified to be the possible sources of these metals, in which crustal sources like soil and road dust due to resuspension by vehicular movement reach indoor microenvironments through outdoor infiltration. Combustion byproducts from scrap and mixed waste burning also reach the indoor spaces in a similar way.

The concentration of iron in indoor sources is multiple. Iron contamination in indoor dust is mostly observed in houses situated near industries and where smelting operations are active. The households in the present study were closer to industries, and numerous other outdoor (traffic emissions) and indoor sources (wear and tear of furniture, Fe-containing paints and varnishes) might have contributed to this trend. In the case of Cd, the concentration obtained in the present study (5.58 mg/kg) was almost similar to the previous studies. Cd contamination in household dust is influenced mainly by exterior sources related to traffic and atmospheric deposition.

3.2 Pollution indices

The contamination factor was computed for various individual elements from various sites and mean CF values of individual metals showed the pattern: Zn > Pb > Cu > Ni > Mn > Cr > Cd > Fe .[2] have studied the contamination factor for heavy metals in dusts (Zn > Cu > Pb > Cr > Cd > Ni), suggesting a pattern similar to this study, with Zn and Pb possessing a high level of contamination compared to other metals. So from the mean CF value for the individual metal, it can be concluded that the Zn and Pb belonged to 'very high contamination' category. Cu belonged to 'considerable contamination' category. Mn, Cr and Ni belonged to 'moderate contamination' category while Fe and Cd belonged to 'low contamination' category.

In case of Pollution load index, except S14 and S17, all sites showed PLI > 1 which implied that 15 out of the 17 sampling sites were contaminated. The order of PLI for various sampling sites showed the pattern: S13 > S16 > S11 > S15 > S9 > S6 > S1 > S5 > S2 > S3 > S12 > S4 > S8 > S7 > S10 > S17 > S14.

3.3 Human health risk assessment

Heavy metals are common constituents of aerosols, and prolonged exposure to these metals affects the internal organs of the human body, such as the liver, kidney, lungs, heart, and brain [3]. The exposure takes place mainly in three forms: inhalation, ingestion, and dermal exposure, by which each heavy metal has its own effect on the human body in terms of chronic exposure. The carcinogenic and non-carcinogenic risk assessments were computed in order to check the short-term human health effects of the population residing in the sampling sites.

3.3.1 Non-Carcinogenic human health risk assessment

The results of the computed hazard quotient for three routes of exposure (children and adults) and overall hazard index (children and adults) revealed that the exposed population residing in the households at varying distances away from the landfill are less prone to short-term health effects from the studied metals. The HI for children in the studied sites possessed the pattern: S5 (0.29) > S13 (0.19) > S16 (0.17) > S11 (0.15) > S6 (0.11) > S15 & S9 (0.10) > S17 (0.09) > S2 & S1 (0.07) > S3 & S4 (0.06) > S8 & S12 (0.05) > S7 & S10 (0.04) > S14 (0.03). In the case of adults, HI followed the pattern: S5 (0.25) > S13 (0.16) > S16 (0.13) > S11 (0.11) > S9 & S6 (0.09) > S15 (0.08) > S17 & S4 (0.07) > S2, S1 & S3 (0.06) > S8 (0.05) > S7 & S12 (0.04) > S10 & S14 (0.03). [2] in his study showed HI < 1, similar to the present study, indicating a normal pattern of heavy metal exposure risk, which further states that there has been no new pattern developed on non-carcinogenic risk. The values of HQ_{inh}, HQ_{ing}, and HQ_{der} for both children and adults are shown in Table 6. Out of the studied pathways of exposure, it concluded that exposure through ingestion was predominant, followed by dermal exposure and inhalation (ingestion > dermal > inhalation) for both children and adults.

Table 6. Values of non-carcinogenic health risk parameters of various heavy metals

HQ	HQ_{inh} (children)	HQ_{inh} (adult)	HQ_{ing} (children)	HQ_{ing} (adult)	HQ_{der} (children)	HQ_{der} (adult)
Mn	0.0099	0.0044	0.0547	0.0294	0.0044	0.0112
Cu	2.56E-06	1.15E-06	0.0458	0.0246	0.0005	0.0012
Cr	0.0019	0.0009	0.3242	0.1740	0.0519	0.1325
Fe	6.97E-06	3.14E-06	0.0005	0.0762	0.0005	0.0012
Cd	2.09E-06	9.40E-07	0.0372	0.0200	0.0119	0.0304
Zn	2.74E-07	1.23E-07	0.0049	0.0026	7.80E-05	0.0002
Ni	1.25E-06	5.63E-07	0.0018	0.0123	0.0003	0.0007
Pb	1.33E-05	5.97E-06	0.2378	0.1276	0.0051	0.0130

3.3.2 Carcinogenic human health risk assessment

Elements such as Cr, Cd, Ni, and Pb were computed for carcinogenic risk assessment, and only the results of Cd provided details regarding CR for all three routes of exposure and the final computation of TCR. The results of the calculated CR (inhalation, ingestion, and dermal exposures) and TCR are provided in Table 7. From Table 7, it is understood that CR_{inh} of Cr, Cd, Ni, and Pb, as well as CR_{der} of Cd, fell in the 'appreciable' category, while CR_{ing} and TCR of Cd fell in the 'acceptable or tolerable' category.

A limited number of studies were only analysed for the carcinogenic effect of heavy metals on humans, in which one study in Chengdu, China [2] showed a category similar to the present study. So the present study concludes that households nearby and away from the landfill have no role in short-term carcinogenic risk to the population residing there.

Table 7. Exposure route values for CR and TCR for heavy metals from the studied sites

Elements	CR_{inh}	CR_{ing}	CR_{der}	TCR
Cr	5.50E-07	-	-	-
Cd	3.16E-09	6.12E-05	6.98E-07	6.19E-05
Ni	5.20E-09	-	-	-
Pb	9.53E-11	-	-	-

4 Conclusion

In order to study the contamination of heavy metals in indoor microenvironments located at varying distances and directions away from the landfill, eight heavy metals, namely Fe, Pb, Cr, Zn, Cu, Mn, Ni, and Cd, were selected based on the elements likely present in waste materials. The results revealed that the concentration of Fe was predominant in the entire sampling location, followed by Mn, Cu, Zn, Cr, Pb, Ni, and Cd. The pollution indices, such as contamination factor and pollution load index, were monitored in order to assess the contamination in the investigated sites and identified 15 sites as contaminated with Zn and Pb as the most contaminated metals. Since Zn and Pb in the study possessed more contamination compared to other elements, and also from the data observed from the previous study, this study could conclude that Zn and Pb could be used as effective markers. Non-carcinogenic and carcinogenic human health risk assessments made on children and adults revealed that the human population in the studied indoor microenvironments was safe and possessed no immediate toxic or harmful effects to the exposed population amidst the fire event. This study further reveals the usage and credibility of pollution indices as well as human health risk assessment for contamination monitoring at a particular site, which can be used in the future for monitoring site-specific contamination as a routine check or during the occurrence of any unexpected events at specific sites.

References

1. V. M. Sreejith, J. R. Thomas, N. K. Khalid, A. Varghese, U. K. Aravind, and C. T. Aravindakumar, *"Implosion–demolitions: impact on the local environment in a tropical humid* Ramsar *site located in South-western India,"* Air Qual Atmos Health, vol. **15**, no. 7, pp. 1243–1261, Jul. (2022).
2. Z. Cheng, L. J. Chen, H.H. Li, J.Q. Lin, Z.B. Yang, Y.X. Yang, X.X. Xu, J.R. Xian, J.R. Shao, X.M. Shu, *"Characteristics and health risk assessment of heavy metals exposure via household dust from urban area in Chengdu, China,"* Sci. Total Environ, vol. **619–620**, pp. 621–629, Apr. (2018).
3. J. Zhang, Z. Chen, D.Shan, Y. Wu, Y. Zhao, C. Li, Y. Shu, X. Linghu, B. Wang, *"Adverse effects of exposure to fine particles and ultrafine particles in the environment on different organs of organisms"* J. Environ. Sci.., vol.**135,** pp. 449-473 (2024).

The Effects of Indoor Air Pollution from Solid Fuel Burning on Neurological and Neuropsychiatric Diseases - A Review of Epidemiological Evidence

Jianing Yang[1], *Minjia* Fan[2], *Linglong* Xu[1*]
[1]University of Queensland, Brisbane, Australia
[2]Qinzhou Center for Disease Prevention and Control, Qinzhou, China

Abstract. Indoor air pollution (IAP) from using unclean fuel has brought huge disease and economic burden, and we aim to analyse the effect of IAP from using solid fuels on neurodevelopmental disorders (NDDs), cognitive impairment and neurodegenerative diseases (NDGDs), as well as mental disorders. We conducted a search of PubMed database from inception to 29th July 2025. We found that the number of studies focusing on the effects of solid fuel burning on NDDs and NDGDs is limited, and the association remains uncertain. However, the evidence regarding cognitive impairment and mental disorders was abundant, and most studies indicated that participants using solid cooking fuels had a higher risk of these diseases. The most important common limitation is that many studies formulaically analyse big health data sets, which leads to problems like residual confounding due to the incomplete information in the datasets. Policy makers of developing countries can prioritize gas as the transitional option from solid fuels to electricity, while developed countries should attach importance to policies related to wood heating. It is crucial to implement policy such as providing financial incentives and offering public education.

Keywords: Indoor air pollution, Solid fuel, Neurodevelopmental disorders, Cognitive impairment, Neurodegenerative diseases, Mental disorders

1 Introduction

It is estimated that there are 2.8 billion people worldwide using solid fuels and simple stoves, and the indoor air pollution (IAP) caused by them has brought a huge disease and economic burden. The use of solid fuels (e.g., wood, coal, crop waste and charcoal) in households causes the emission of several indoor air pollutants, such as particulate matter (PM), sulphur oxides, carbon monoxide (CO), heavy metals and volatile organic

* Corresponding author: nataliexu00@gmail.com

© The Authors, published by EDP Sciences. This is an open access article distributed under the terms of the Creative Commons Attribution License 4.0 (https://creativecommons.org/licenses/by/4.0/).

compounds (VOCs). These pollutants can adversely affect human health, causing diseases include stroke, ischemic heart disease, lower respiratory infections and chronic obstructive pulmonary disease (COPD). The global loss of healthy life years due to IAP was estimated to be 86 million in 2019 [1], and the global welfare losses from household air pollution was estimated to be 1.516 trillion dollars in 2013 [2].

Humans spend about 90% of their time indoors [1], and indoor air quality is a significant factor of indoor environment quality. In addition, neurodevelopmental disorders (NDDs), cognitive impairment (CI) and neurodegenerative diseases (NDGDs), as well as mental disorders have imposed huge health and economic burden. Thus, it is important to analyse the impact of IAP on these neurological and neuropsychiatric diseases. Here, CI and NDGDs were discussed together because NDGDs has a long preclinical period and CI is closely associated with NDGDs such as Alzheimer's disease (AD) and Parkinson disease (PD) [3, 4]. We found that the number of studies related to NDGDs was limited, while literature focusing on CI was abundant. Therefore, the impact of IAP on CI may also imply the impact on NDGDs. Previous research has reported that air pollution can exert neurotoxicity across life span. Mother exposed to air pollutants (e.g., particulate matter, sulfur and iron) can affect placenta and the development of fetal brain. After birth, air pollutants can enter the blood circulation after being inhaled and then travel into the brain. The shared mechanisms by which air pollutants affect NDDs and NDGDs may include inflammation and activation of microglia, altered glutamatergic function and mitochondrial dysfunction. These lead to shared outcomes of white matter damage and ventriculomegaly [5]. In addition, air pollutant can also affect mental health through neurostructural and neurofunctional changes (e.g., inflammation and oxidative stress, neurotransmitter change) in hippocampus or multiple brain regions [6]. Furthermore, epidemiological studies also explored the effect of IAP from solid fuel use on NDDs, CI and NDGDs, as well as mental disorders. Thus, we review the epidemiological evidence from PubMed, as well as provide suggestions for future research and policy formulation.

We performed search in PubMed from database inception to 29th July, 2025. The selection criteria are as follows. The exposure needs to be related to IAP due to solid fuel use, and the outcomes must be NDDs, NDGDs and mental disorders. The searching terms include "indoor air pollution", "autism spectrum disorder", "ADHD", "intellectual disability", "Alzheimer disease", "Parkinson disease", "cognitive dysfunction", "dementia", "depression" and "angst". We used the MeSH database to identify the terms and the synonym, and our search strategy incorporated MeSH terms and the possible expressions to ensure the comprehensive coverage of the literature. The search was limited to peer-reviewed epidemiological studies published in English. A total of 632 articles (81 on NDDs, 341 on CI and NDGDs, 210 on mental disorders) were found, of which 174 articles (23 on NDDs, 92 on CI and NDGDs, 59 on mental disorders) were selected for full-text reading. Finally, 57 studies (5 on NDDs, 31 on CI and NDGDs, 21 on mental disorders) met our selection criteria. However, we found that many articles focusing on cognitive impairment and mental disorders have similar research designs and conclusions, so we discussed the common findings of these studies and only selected some valuable articles for detailed discussion. The shared limitations were further summarized in section 5.

2 Neurodevelopmental Disorders

Neurodevelopmental disorders (NDDs) often manifest during developmental stages, typically manifest early in life (before entering elementary school) and are characterized by developmental deficiencies that impact the development of brain and nervous system, which in turn impacts individual, social, academic, or professional functioning. These diseases have brought huge health and economic burden. According to the survey, the incidence of ASD and ADHD in children ages 0-14 worldwide was 1,163,706.03 and 4,111,621.02 in 2021 respectively [7]. In 2019 alone, the total socioeconomic cost of ADHD in Australia came to about A$20.42 billion, of which A$10.19 billion was due to lost productivity. And the annual financial cost of ASD in the US is projected to reach $461 billion by 2025 [8, 9]. We selected three common NDDs: autism spectrum disorder (ASD), intellectual disability, attention deficit hyperactivity disorder (ADHD), but we did not find studies related to intellectual disability. Therefore, a total of 5 studies focusing on ADHD and ASD were discussed in this section.

The characteristics of the selected papers were shown in **Table 1**. In terms of ADHD, there is a study in Guangzhou which used a cross-sectional approach. It included 7,495 schoolchildren aged 6 to 12 and found a significant positive association between exposure to cooking oil smoke and a higher incidence of ADHD symptoms [10]. In addition to elementary school children, those affected by indoor air pollution may also include pregnant and preschool children. A prospective cohort study involving 42,983 mothers conducted between 2015 and 2017 concluded that maternal exposure to cooking fumes during pregnancy was significantly positively associated with the risk of hyperactive behaviour in their children at age three [11]. Moreover, there is also a positive correlation between IAP and autism-like behaviours. A cross-sectional study conducted in Shenzhen concluded that there was a significant positive dose-response relationship between the amount of cooking oil fumes (COF) exposure or cooking frequency during pregnancy and the likelihood of autism-like behaviours in preschool children [12]. A randomized controlled trial involving 540 pregnant women in Ulaanbaatar, Mongolia, conducted between 2014 and 2015 also concluded that increased indoor $PM_{2.5}$ concentrations during pregnancy caused by coal burning in household heating stoves were associated with higher scores on autism-related behaviours in children [13]. However, a cohort study in southern Sweden showed that although $PM_{2.5}$ exposure from small building heating was also associated with ASD, the association was less significant in the fully adjusted mode [14]. This may be due to the different backgrounds of different countries regarding this fuel, involving socio-economic factors, production requirements, lifestyle adjustments, etc [15].

Table 1 Characteristics of the selected papers on NDDs

Ref.	Location	Study type	Key findings
[10]	Guangzhou, China	C-S	Participants with COFs exposure had a higher ADHD index and higher odds of ADHD symptoms. (OR=2.33, 95%Cl: 1.48, 3.66; OR=1.36, 95%CI: 1.04, 1.74)

Ref.	Location	Study type	Key findings
[11]	Shenzhen, China	Cohort	In preschoolers, those with high household air pollutants exposure had a higher risk of hyperactive behaviours.
[12]	Shenzhen, China	C-S	There is a positive association between COFs exposure and autistic-like behaviours in child, with the AORs for "often" was 1.66 (95%CI: 1.42, 1.95) and for "always" was 2.29 (95%CI: 1.96, 2.67).
[13]	Ulaanbaatar, Mongolia	RCT	Over the course of the pregnancy, a 9.6 μg/m^3 increase in indoor $PM_{2.5}$ concentration was linked to a mean SRS-2 T-score increase of 1.8 units (95 %CI: 0.3, 3.2).
[14]	Scania	Cohort	$PM_{2.5}$ exposure from small residential heating was associated with ASD in the partially adjusted model (OR=1.22; 95%CI:1.08, 1.37), while in fully adjusted model, OR= 1.12 (95%CI: 0.94, 1.34)

Note: ADHD, attention deficit hyperactivity disorder; AORs, adjusted odd ratios; ASD, autism spectrum disorder; CI, confidence interval; COFs, cooking oil fumes; C-S, cross-sectional study; OR, Odds Ratio; PM, particulate matter; RCT, Randomized Controlled Trial; Ref., References; SRS-2 T-score, Social Responsiveness Scale, 2nd edition

Therefore, the number and focus of studies on the effects of IAP from indoor combustion or heating on NDDs are limited. Current studies focus primarily on common products produced by indoor combustion, such as $PM_{2.5}$. Most studies have shown its negative effects on children's attention and cognitive function. In addition to $PM_{2.5}$, other pollutants produced by indoor solid fuel use vary depending on the fuel [16]. Their impact on children's intellectual development also warrants more extensive research, and the study region should be expanded beyond China and Europe.

3 Cognitive impairment and neurodegenerative diseases

3.1 Cognitive impairment

Cognitive impairment refers to dysfunction of multiple aspects of cognition. A person with cognitive impairment often experiences difficulty in learning, remembering, focusing and decision making. Mild cognitive impairment (MCI) is common in the elderly and often regarded as the phase that lies between normal aging and dementia [3]. It was estimated to have a global prevalence of 19.7% and carry a great economic burden. For example, a study conducted in Germany estimated that the medical and formal care costs for MCI individuals were elevated by 85% compared to the healthy controls, and the mean healthcare costs per year were 8,795€ with a 95%CI of 6,200 to 11,391€ [17]. In addition, individuals in the United States progressing from MCI to Alzheimer disease and related dementia disorders experienced substantially higher medical costs than those with stable MCI [18], so it is important to early identify MCI and manage potential disease influencing factors. The main finding was that participants exposed to IAP from unclean fuel use had a higher risk of cognitive impairment, and the specific findings were discussed from the perspectives of exposure, outcome as well as potential influencing factors in this section.

The details of the selected articles were presented in Table 2. Regarding exposure, solid fuel use for cooking had adverse effects on cognition, but there is uncertainty in terms of heating and lighting. For example, a study showed that the association between solid heating fuel use and total cognitive score was not statistically significant during the follow-up, but the cross-sectional results in the same article was statistically significant [19]. In addition, two articles demonstrated low effect size in terms of solid heating fuel use [20,21]. This might be related to differences in exposure intensity and duration. Exposure to indoor air pollutants from cooking is more likely to be regular and long-term, resulting in a larger cumulative exposure. Heating tends to be seasonal (only in winter) and exposure to the generated pollutants is relatively low [19]. It is also worth noting that there is a study conducted in Ireland [21], there might be differences in socioeconomic factors (e.g., educational levels, culture and management of chronic diseases) between developed and developing countries, resulting in smaller effect size observed in study conducted in developed countries [15]. Further multi-national research is needed to determine the impact of solid heating fuel use on cognitive function. Moreover, an article showed that use coal for cooking can elevate the risk of cognitive decline [22], while another found such association in wood use but not coal use [23]. The potential reason is that China promoted improved coal-fuelled stoves in recent years, and such stoves was reported to be able to reduce the emission of coal burning pollutants [24]. The former study analysed data from 1997 to 2015 [22] and the latter used data from China Health and Retirement Longitudinal Study (CHARLS) wave 2011, 2013 and 2015 [23]. Participants from earlier cohort may be exposed to more indoor air pollutants from coal use because of using traditional stoves, and thus had a significant effect size. Furthermore, some literature also demonstrated the benefit of changing fuels from solid to clean [19, 22, 25, 26], especially switching cooking fuels. These findings indicated the great benefits of switching to improved stoves and cleaner fuels, which can not only mitigate the risk of cognitive decline among the elderly, and is potentially beneficial for the recovery of cognitive function because MCI can be slowed or partially reversed [27]. Finally, two articles measured exposure to specific pollutant and found that black carbon, particulate matter, NO_2 and O_3 can adversely affect cognitive health [28,29]. Although the number of participants were relatively limited (120 participants and 401 participants), but these research addressed the greatest uncertainty in most literature, which is the lack of significant information (e.g., temperature, humidity, exposure duration and frequency) and recall bias caused by measuring exposure through questionnaires. The design of these articles holds significant reference value for future research, and the details will be discussed in section 5.

In terms of the outcome, the literature we found explored the impact to specific domains of cognitive health in addition to the overall cognitive function. However, there is significant variation in the domains because there is not a universally accepted tool for cognitive function screening. The commonly explored domains were episodic memory, visuo-spatial function and attention. And the most frequently used tools were Mini-Mental State Examination (MMSE). In recent years, many systematic reviews were conducted to explore the sensitivity and specificity of different screening tools, and each tool has advantages and applicable scenarios [30,31]. It is necessary to develop a widely applicable screening tool, and we recommend using multiple tools in studies focusing on the association between environmental pollutant and cognitive function. In addition, some research investigated the effect of solid fuel use on multimorbidity. For example, a study conducted in 2024 employed data from CHARLS and identified physical-psychological-cognitive multimorbidity (PPC-multimorbidity) as the co-occurrence of physical illness (stroke, heart disease, hypertension, diabetes, lung disease, arthritis or cancer), depression and cognitive impairment. Their results suggested the adverse effects of using solid fuels and the benefits of changing to cleaner fuel [32]. Similarly, another research conducted in 2025 also analysed data from CHARLS and identified cognitive frailty (CF) as the co-occurrence of cognitive impairment and physical frailty. The results showed that participants using solid fuel for cooking or heating had a higher risk of CF, and those who changed from solid to clean fuel for cooking showed a decreased risk of CF [33]. These articles imply future research directions may be related to multimorbidity. It has an estimated global prevalence at 51% among the elderly, and the complex interactions among diseases also affect the vulnerability of the elderly [34]. Future research needs to dive deeper into multimorbidity related to cognitive impairments and the interaction among diseases.

There were also some influencing factors that may affect the association between IAP and cognitive health. The most frequent studied factor was gender, and the studies of developing countries demonstrated that females were more vulnerable. One explanation is that women in developing countries often bear responsibility of household chores such as cooking. Compared with males, their exposure from using solid fuel is characterized by longer duration and more regular frequency, and thus more exposure to indoor air pollutant [1]. Another explanation is that females might be more vulnerable to air pollutants such as particulate matter. $PM_{2.5}$ is an important product of indoor solid fuel combustion, and the neuroinflammation it induces serves as the potential mechanism of cognitive decline [35]. Previous mechanistic studies have demonstrated that females had greater susceptibility to neuroinflammation. For example, among rats exposed to $PM_{2.5}$ for 16 months, only female subjects exhibit elevated levels of interleukin-6 in the hippocampus. And the level of tumour necrosis factor-α increased in most brain regions in female but solely heightened in male hypothalamus [36]. However, it is worth noting that the study conducted in Ireland found contrary results [21]. The possible reason is that the focus of this article is on heating not cooking, and it is the only study conducted in developed European countries. There are sociocultural differences between different countries, which may result in relatively low exposure of indoor air pollutants for Irish women. In addition to gender, many relevant factors that may affect the association were also investigated, such as age, educational level, residence, lifestyle (e.g., physical activities, smoking history, drinking history), health condition (e.g., cardiovascular diseases, diabetes, depression, sleeping disorders), self-care ability and ventilation facilities. However, the research we discovered often applied logistic regression and regarded these factors as independent. Few studies have explored the interactions among these factors and their impact on cognition. One study is with relatively certain reference value regarding methodology [37]. The authors applied Lasso regression to select the potential factors and Bayesian networks to explore the dependencies among factors. And they found that indoor air pollution, age and educational level directly affect cognitive impairment in the older adults. Residence (rural/urban) can indirectly affect cognitive health through educational level or indoor air pollution, while the indirect influence of marital status was through age. In the future, more machine learning technologies should be applied to explore the interactions not only among factors but also between influencing factors and multimorbidity.

Table 2 Characteristics of the selected papers on cognitive impairment

Ref.	Location	Study Type	Key Findings
[19]	China	C-S and cohort	In the cross-sectional analyses, participants using solid fuel for cooking (β = −0.65; 95%CI: −0.85, −0.45) and heating (β = −0.39; 95%CI: −0.63, −0.16) had lower total scores of cognitive functioning. In the follow-up study, participants using solid cooking fuel had lower total scores of cognitive functioning (β = −0.24, 95%CI: −0.45, −0.04). Participants who switched from solid to cleaner cooking fuel had a lower degree of cognitive impairments.
[20]	China	Cohort	Participants using solid fuel for cooking (OR=1.11, 95%CI: 1.01, 1.23) and heating (OR=1.15, 95%CI: 1.01, 1.31) had a higher risk of MCI. Healthier lifestyle, especially healthier BMI, can reduce the risk.
[21]	Ireland.	Cohort	Participants using solid heating fuel had a higher risk of cognitive decline (b = −0.22, 95%CI: -0.37, −0.08, p = 0.003). There were statistically significant association observed in males, but not in females.
[22]	12 provinces in China	Cohort	The use of unclean fuel for cooking was negatively associated with global cognitive function (P<0.05), but the association between the use of unclean fuel for lighting and global cognitive function was not statistically significant (P>0.05). Participants switching from unclean to clean cooking fuel had a better global cognitive function. Hypertension was the potential mediator. Those who were female, ≥65 years, living in rural area and using coal for cooking were vulnerable group.
[23]	China	C-S	There was sex difference in the association between solid cooking fuel and cognitive impairment (P=0.008). Females were more vulnerable than males. Wood use was associated with cognitive decline in cooking and heating, and no such association was found in coal use.

Ref.	Location	Study Type	Key Findings
[25]	China and Mexico	Cohort	Participants using solid cooking fuel had a higher risk of cognitive decline in the cohort of CFPS, CHARLS and MHAS. Those who switch from solid to clean fuel showed a decreased risk of cognitive decline. Physical activity, diabetes, hypertension dyslipidemia and HDL-C were potential mediators.
[26]	China	Cohort	Participants using biomass cooking fuel had a higher risk of cognitive impairment and cognitive decline. Those who switched from biomass to clean cooking fuels had a lower risk of cognitive impairment and cognitive decline. The use of ventilation facilities, living with family members and having a regular exercise pattern can reduce the risk.
[28]	Two rural villages of Shanxi Province, China	C-S	There were positive associations between indoor PM_1, $PM_{2.5}$, PM_{10}, NO_2, O_3 exposure and the risk of cognitive impairment (for every 1μg/m^3 increase in PM_1, $PM_{2.5}$, PM_{10}, NO_2, and O_3, OR:1.070 (1.040–1.100), OR: 1.038 (1.022–1.054), OR: 1.028 (1.016–1.039), OR: 1.045 (1.007–1.084), OR: 1.021 (1.010–1.033), respectively). The female group is more vulnerable than male group.
[29]	Northern China	C-S	Overall cognitions were negatively associated with fine particulate matter exposure (-0.13, 95%CI: -0.22, -0.04) and black carbon exposure (-0.10. 95%CI: -0.19, -0.01). Attention was the cognitive domain that had the strongest association with the exposure.
[37]	China	C-S	The direct influencing factors of cognitive impairment in the older adults were age, educational level and IAP. Marital status and residence (rural/urban) were indirect influencing factors.

Note: BMI, body mass index; CFPS, the China Family Panel Studies; CHARLS, the China Health and Retirement Longitudinal Study; CI, confidence interval; C-S, cross-sectional study; HDL-C, high-density lipoprotein cholesterol; MCI, mild cognitive impairment; MHAS, the Mexican Health and Aging Study; OR, odds ratio; PM, particulate matter; Ref., references.

There were three systematic review and meta-analysis that investigate the relationship between IAP and cognition [38-40]. Their results all support that participants using solid fuel had a higher risk of cognitive impairment and switching to cleaner fuels can lower the risk. However, these studies shared similar limitations such as the small number of the included studies and high heterogeneity. Future epidemiological research should be conducted in more countries and focusing on specific indoor air pollutants, and meta-analysis should apply methods like Hartung-Knapp-Sidik-Jonkman (HKSJ) adjustment to deal with the substantial heterogeneity.

3.2 Neurodegenerative disorder

Mild cognitive impairment is also called as "preclinical dementia", it progresses to Alzheimer disease (AD) at the yearly rate of 10%-15% [3]. AD is a common neurodegenerative disorder (NDGD) and the most typical form of dementia, accounting for about 60% to 70% of dementia cases. From 1991 to 2021, the global prevalence of Alzheimer disease and other dementias (ADRD) increased by approximately 1.6 times among individuals aged 65 years and above, and the worldwide economic burden in 2019 was about 2.8 trillion dollars [41]. Parkinson disease (PD) is another NDGD, which is characterized by a maximal tremor at rest, mask-like facial expression, slow voluntary movements, rigidity, retropulsion and stooped posture. It was estimated that over 8.5 million people suffered from PD worldwide in 2019 [42] and the financial burden in the United States reached 51.9 billion dollars in 2017 [43]. In this review, we retrieved a total of 226 articles from PubMed, 6 met our selection criteria and were discussed in the following text. Noteworthy, we did not find any article directly investigating the association between AD and IAP, so we discussed studies on all-cause dementia to explore the potential effect of IAP on AD. Table 3 displayed the features of the selected studies.

Regarding the association between dementia and IAP, we found 5 studies, of which 3 was conducted in China [44-46]. They all analysed data from CHARLS and found that IAP increases the risk of dementia. Depression and accelerated biological age can influence the association. However, the remaining 2 articles had different findings [47,48]. They were both conducted in northern Sweden and the effect of IAP on dementia was uncertain. One study found that the association between dementia risk and 1μg/m^3 increase in $PM_{2.5}$ exposure from indoor wood burning was not statistically significant [47], while another reported that $PM_{2.5}$ from indoor wood combustion even might have a protective effect on dementia [48]. One potential reason is that these two studies were both conducted in the developed country, which has better socioeconomic status. As mentioned in section 3.1, factors such as educational level, lifestyle and health condition can affect the association between IAP and cognitive health. These factors may be more advantaged among the elderly in Sweden than those in China [15]. These brain-protective advantages may statistically mask the adverse effect of indoor air pollutants on dementia, which suggests the importance of studying more influencing factors and adjusting the residual confounders. Another reason is that using different fuels produce different pollutants with varying toxicities. The two studies in Sweden both investigated the $PM_{2.5}$ exposure from indoor wood burning [47, 48], while the other 3 studies did not mention the fuel type [44-46]. Taking $PM_{2.5}$ as an example, the toxicity of $PM_{2.5}$ from burning coal and burning biomass is different [16]. In terms of PD, the epidemiological evidence is very limited, and we only found 1 cross-sectional study conducted in Peru [49]. The study showed that the prevalence of PD is higher among those using wood cooking fuels. But the study used prevalence ratio as effect estimate, which does not capture incidence and not estimate disease risk.

In summary, the impact of IAP on dementia, AD and PD remains uncertain. The number of epidemiological research is very limited and further multinational investigations is needed to enrich our understanding.

Table 3 Characteristics of the selected papers on NDGDs

Ref.	Location	Study Type	Key Findings
[44]	China	C-S and retrosp-ective longitudinal analyses	In the cross-sectional study, participants using solid cooking fuel had a higher risk of dementia (OR = 1.22, 95%CI: 1.02, 1.47). Similar results were observed in the retrospective longitudinal analysis (OR = 1.44, 95%CI:1.06, 1.96). Depression partially affects the correlation between solid cooking fuel use and the risk of dementia.
[45]	China	Cohort	Participants using solid fuels for cooking or heating had a higher risk of dementia. The vulnerable group were females and those living in rural areas.
[46]	China	Cohort	Compared with those consistently using cleaner fuels, participants consistently using solid fuels had a higher risk of dementia (HR = 1.32, 95 % CI: 1.06, 1.64). Accelerated biological age partially affect the association between consistent solid fuel use and dementia.
[47]	Umeå, Northern Sweden.	Cohort	For every 1μg/m^3 increase in $PM_{2.5}$ exposure from indoor wood burning, the increasing risk of dementia was not statistically significant (HR=1.55, 95%CI: 1.00–2.41, p = 0.05).
[48]	Northern Sweden	Cohort	Exposure to $PM_{2.5}$ from indoor wood burning was negatively associated with dementia risk.
[49]	Northern Peru	C-S	Participants using wood as cooking fuels had a higher prevalence of Parkinson disease (PR = 3.45; 95%CI: 1.25; 9.48).

Note: CI, confidence interval; C-S, cross-sectional study; HR, hazard ratio; OR, odds ratio; PM, particulate matter; PR, prevalence ratio; Ref., references.

4 Mental disorders

Depression and anxiety are two most typical mental disorders. Depressive disorder (also known as depression) refers to prolonged low mood or loss of happiness or interest in activities, which are caused by the complex interaction of social, psychological and biological factors [50]; anxiety patients feel excessive panic or concern about certain specific situations [51]. These two mental disorders have become a major burden on global public health. According to WHO, approximately 280 million people worldwide suffer from depression, and up to 2019, 301 million people were suffered from anxiety disorder, the global financial burden of depression and anxiety is US$ 1 trillion each year [50, 51]. Table 4 showed the characteristics of the selected articles.

In terms of study design, we found that solid fuel exposure was mainly defined through self-reported fuel types in questionnaires, while depressive and anxiety symptoms were primarily assessed using screening tools such as Centre for

Epidemiological Studies Depression scale (CES-D) (12 studies), Patient Health Questionnaire (PHQ) (4 studies), and Generalized Anxiety Disorder (GAD) (5 studies). Overall, the key findings appear high consistency, indicating a positive association between solid fuel use and symptoms of anxiety and depression. However, when solid fuels were used only for heating purposes, no significant association was observed [52]. Vulnerable groups such as women, the elderly, rural residents, low-income populations, and individuals with low educational levels were found to be at a higher risk.

One of the earliest exploratory studies in this field was conducted in 2012 [53], despite its significant limitations, we consider that this paper remains highly valuable for research purposes: its cross-sectional design precludes the establishment of causality, and residual confounders such as unmeasured social stress may partially explain the effect size; the study was restricted to rural women of childbearing age in eastern India, and the measurement of depression outcomes relied on self-report screening tools. Nevertheless, the finding was moderate effect size and it still provides an important hypothesis foundation for subsequent research. Unlike recent papers, its data were recruited from volunteer rather than secondary analysis of large-scale cohort study data, and it preliminary explored depression related biomarker associations. We believe the core value of this paper lies in proposing hypotheses and paving the way for future study design.

Table 4 Characteristics of the selected papers on mental disorders

Ref.	Location	Study type	Key findings
[52]	China	Cohort	The use of solid fuels for cooking was associated with an increased risk of depression (HR=1.12, 95%CI: 1.02, 1.24), while heating alone showed no significant association. Combined use for heating and cooking was associated with higher risk (HR=1.15, 95%CI: 1.01, 1.31).
[53]	Eastern India, rural areas	C-S	The use of solid fuels significantly affected the risk of depression (OR=1.67, 95%CI: 1.18, 2.95).
[54]	Henan, China	C-S	Solid fuel use was associated with increased risks of depression (OR=1.237, 95%CI: 1.041, 1.469) and anxiety (OR=1.384, 95%CI: 1.153, 1.662). Longer cooking time was linked to higher risks of depression and anxiety, particularly in households without ventilation.
[55]	Henan, China	Cohort	The use of solid fuels was associated with an increased risk of depression (OR=1.46, 95%CI: 1.12, 1.89) and anxiety (OR=1.45, 95%CI: 1.10, 1.89).
[56]	Oaxaca, Mexico	C-S	The study showed higher risk of depression among women, participants aged 60 and above, those living in rural areas, and people using wood or charcoal for cooking.

Note: CI, confidence interval; C-S, cross-sectional study; HR, hazard ratio; OR, odds ratio; Ref., references.

The common limitation is the insufficiency in controlling confounding factors. Depression and anxiety are mental disorders caused by multiple factors that involve complex interactions among genetic, environmental, and social levels. We identified several key confounding factors, which should be incorporated into future research: socioeconomic status (SES), living environment, ventilation conditions, chronic diseases, and exposure to second-hand smoking. Among these factors, ventilation condition plays an important role. Although most of the studies we found accounted for ventilation, many did so insufficiently by simply asking participants about its presence without distinguishing between mechanical and natural ventilation. However, some papers still partially explored the moderating role of ventilation conditions. For instance, one study presented that mechanical ventilation may eliminate the risk of depression associated with solid fuel use and prolonged cooking (OR reduced from 1.54 to 1.02), whereas natural ventilation does not provide such protective effects [54]. Another prospective cohort study found that 46% of the depression and anxiety risk could be attributed to the synergistic effect of using solid fuels and natural ventilation [55]. This finding implies that ventilation is not only an important confounding factor, but also an intervenable effect-modifying factor.

We also found that most papers reached the result of small effect size. Noteworthy, one study reported an extraordinary OR value of 12.192 (95%CI: 12.064, 12.321, $p < 0.001$) in the subgroup analysis of the elderly population [56]. The authors did not directly explain the cause of this abnormal value, but implied that it may be due to the elderly population were exposed for a longer time, hence increasing the risk. One possible reason is that the study ignored some key confounding factors in the older population, such as chronic disease conditions and disabilities. Another reason is that the categorization of the elderly was inadequate in this study. In the subgroup analysis, the older population was simply categorized as those aged ≥60 years. But relatively younger elderly people tend to have more resistance than the older elderly. Older elderly individuals may have a higher risk of depression due to health decline and longer exposure time [57], which may lead to the effect size of this subgroup were amplified. In addition, the possible reasons for the small effect sizes are as follows. Almost all papers measured anxiety and depression outcomes through screening tools such as CES-D and GAD, which may result in non-differential misclassification bias. Since the possibility of each participant being misclassified is the same, this will bias the effect estimate towards the null, resulting in a systematic underestimation of the association strength. Further studies should verify the results of screening tools through a more rigorous outcome definition (e.g., International Classification of Diseases, ICD codes) to lower the non-differential misclassification bias. If screening tools are the only way to measure outcomes, a quantitative bias analysis should be included to measure the magnitude of underestimation of the effect size [58].

In conclusion, despite the association between household solid fuel or biomass fuel use and depression and anxiety is notable, current studies are designed based on secondary analysis of large-scale cohort database, and the results of these studies should be interpreted cautiously. Future research should include biomarker detection in the study design, longer-term cohort studies should also be considered to further enhance the credibility of the evidence.

5 Research limitations

Although the number of studies focusing on cognitive impairment and mental health disorders is relatively large, we need to interpret our findings with caution. This is because these articles suffered from the issue of formulaically analyse big health data sets, especially in the field of cognitive impairment and mental health disorders. This has been criticized many times this year [59,60]. Big health data set such as CHARLS and Longitudinal aging study in India (LASI) contains a wealth of health-related measurements and involves a large number of participants, but it often leads to incomplete information when it comes to analysing specific health problems.

First, the common limitation of these studies is the self-reported exposure data. The studies often depend on the investigation of fuel type and stove use to determine exposure, lacking crucial data (e.g., exposure frequency, ventilation, humidity and temperature) and inevitably induce recall bias. Contents that remain uncertain include the interaction effect of household fuel use and outdoor air pollution, as well as the impact of mixed using clean and solid fuel during fuel transition. Another common limitation is that there might be residual confounding. Although previous studies indicated that factors such as health conditions and self-rated health can affect the association between IAP and neurological diseases, some big health data set do not measure this type of information. Regarding outcome determination, many studies used screening tools (e.g., CES-D and the abbreviated Community Screening Instrument for Dementia) rather than diagnostic criteria, resulting in misclassification. However, some studies have also attempted to address these deficiencies and these study designs have great reference value. For example, one study provided detailed descriptions of the methods for measuring personal exposure to specific pollutants and checking the compliance of the participants [29]. Another research not only measured specific indoor air pollutant exposure but also analysed microbiota from the fecal sample. The authors found that cognitive impairment was associated with air particulate exposure through microbiota gut-brain axis [28]. In addition, some researchers performed brain MRI to 994 participants and found females using polluting cooking technology had lower hippocampus volumes [61]. Furthermore, the number of the studies with a cross-sectional design is greater than that of longitudinal design. For example, in the studies focusing on cognitive impairment, 1 article presented both cross-sectional and follow-up results, 15 articles were cross-sectional studies and 9 were cohort studies. In addition, among the 21 studies focusing on mental disorders, only 4 utilized a cohort design. But cross-sectional survey can only identify association not causality and there might also be reverse causation. For instance, participants with lower cognitive function or mental health problems may be less likely to choose or switch to use clean fuel. Lastly, we also found that most research was carried out in developing countries with a focus of solid cooking fuel use, especially in China and India. Among the 25 research on cognitive impairment, 15 were conducted in China and 9 in India. Similarly, of the studies on mental disorders, 16 were carried out in China and 2 in India. But developed world also face the problem of using solid heating fuel, and more research should be conducted exploring its impact on neurological disease.

In summary, present studies provide information for the design of future cohorts specifically exploring this research problem, and scholars should pay attention not to produce similar articles based on big health data sets.

6 Policy analysis

According to WHO, 1.8 billion people worldwide will be unable to access clean fuels and technologies by 2030 if there are not policy intervention, so policy reform is crucial [1]. In the following texts, we analyse the policy approaches and current challenges for household energy and stoves in developing and developed countries.

Developing countries have made efforts in policy implementation and achieved success. In China, efforts originated in the early 1980s. The National Improved Stoves Programs (NISP) introduced 180 million improved stoves with chimneys [62]. From 2017 to 2021, the Clean Winter Heating Plan was implemented in northern China and encompassed more than 200 million residents living in rural areas [63]. However, wealthier areas and households transferred to clean fuels and stoves at a faster pace, while the most disadvantaged areas even saw a regression from clean fuels back to solid fuels. Similar problems were observed in India. Recent research has shown that despite the Pradhan Mantri Ujjwala Yojana (PMUY) increasing the connection rate of liquefied petroleum gas (LPG) to over 95%, nearly 85% of households mixed use LPG and solid fuels [64]. The factors that affect acceptability include the affordability of clean fuels and improved stoves, as well as cultural preferences (e.g., Traditional meals cooked with solid fuels tastes better). Therefore, the policy priority in developing countries should be on long-term financial supports on fuel supply as well as awareness campaigns such as community education programs [65].

The efforts of developed countries mainly focus on the wood burning appliances. The U.S. Environmental Protection Agency (EPA) sets New Source Performance Standards (NSPS), which are the federal standards for residential wood heaters. Manufactures of household wood heaters must be certified by EPA before they can sell the heaters in the U.S. In addition, there is 26% Federal tax credit for wood heaters and EPA provides guidance for community-level changeout programs for wood burning. However, a report pointed out the challenges in heater testing and standard enforcement, which result in some of the wood heater sold in the U.S. emitting air pollutants that exceed the standards [66]. The boldest policy is in the Australian Capital Territory (ACT), with the Cabinet's pledge to eliminate wood heaters by 2045. Although this policy seems to have encountered political resistance recently [67], the repetition and debate of the "phase-out agreement" have attracted political and public attention, which are also a manifestation of progress. In addition to policies on wood burning appliances, developed countries tried to formulate and implement laws and regulations on switching gas to electric stoves. For example, the New York State in the U.S. updated its building code this year, which bans using gas equipment in the majority of new buildings [68]. From the perspective of the health impact of fuel, research showed that using gas increased the risk of COPD and pneumonia comparing to using electricity [69]. However, researchers from the same study also reported that using gas can reduce various health risks compared with using solid fuels. Therefore, it is appropriate for developing countries to use gas as a transitional fuel to reduce the disease burden.

In short, both developing and developed countries have made efforts in policy related to the switch of household cleaner fuels. The main challenges include standard enforcement, political resistance, energy poverty and social inequities. Policy makers should pay attention to efficient economic subsidy measures, educational materials tailored to specific cultural backgrounds, and stakeholder (e.g., expert panels, public health advocators, traditional energy industry) communication. These policies are beneficial for improving both indoor and outdoor air quality, as well as reducing related health risks.

7 Conclusion

Overall, we conclude that the impact of IAP from using solid fuels on NDDs and NDGDs remains uncertain due to the limited research number, and those using unclean fuel generally had a higher risk of cognitive impairment and mental disorders. The common limitation of the studies we found is the formulaic analysis of big health datasets, and the results should be interpreted with caution because the analysis was based on incomplete information. Long-term cohorts, specific indoor air pollutant exposure measurements and biomarker detection should be prioritized in the future research design. Developing countries should promote phasing out solid fuel and the renewal of improved stoves, and the developed countries should pay more attention to policies on wood heating. Efficient economic subsidy measures and public education adapted to specific cultural backgrounds are recommended policy instruments for both developing and developed countries.

References

1. WHO (2024) Household air pollution
2. World Bank (2016) The Cost of Air Pollution: Strengthening the Economic Case for Action (Vol. 1 of 2)
3. Parnetti L, Chipi E, Salvadori N, et al (2019) Prevalence and risk of progression of preclinical Alzheimer's disease stages: a systematic review and meta-analysis. Alzheimer S Research & Therapy 11:.
4. Degirmenci Y, Angelopoulou E, Georgakopoulou VE, Bougea A (2023) Cognitive Impairment in Parkinson's Disease: An updated overview focusing on emerging pharmaceutical treatment approaches. Medicina 59:1756.
5. Cory-Slechta DA, Merrill A, Sobolewski M (2022) Air Pollution–Related neurotoxicity across the life span. The Annual Review of Pharmacology and Toxicology 63:143–163.
6. Zundel CG, Ryan P, Brokamp C, et al (2022) Air pollution, depressive and anxiety disorders, and brain effects: A systematic review. NeuroToxicology 93:272–300.
7. Liu Z, Liu Y, Yan L, Li Q (2025) Trends in the global burden of disease for mental disorders among children aged 0-14 years, 1990-2021. Journal of Affective Disorders.
8. Deloitte (2023) The social and economic costs of ADHD in Australia | Deloitte Australia | Deloitte Access Economics, Healthcare, Economics

9. Leigh JP, Du J (2015) Brief report: Forecasting the economic burden of autism in 2015 and 2025 in the United States. Journal of Autism and Developmental Disorders 45:4135–4139.
10. Chen Y-C, Gui Z-H, Bao W-W, et al (2022) Chronic exposure to indoor air pollutants in association with attention-deficit/hyperactivity disorder symptoms in Chinese schoolchildren: A cross-sectional study. NeuroToxicology 94:182–190.
11. Fang X-Y, Strodl E, Liu B-Q, et al (2019) Association between prenatal exposure to household inhalants exposure and ADHD-like behaviors at around 3 years of age: Findings from Shenzhen Longhua Child Cohort Study. Environmental Research 177:108612.
12. Yang J-H, Strodl E, Wu C-A, et al (2022) Maternal exposure to cooking oil fumes during pregnancy and autistic-like behaviors in Chinese preschoolers. Environmental Science and Pollution Research 29:74688–74698.
13. Enkhbat U, Gombojav E, Banzrai C, et al (2022) Portable HEPA filter air cleaner use during pregnancy and children's autistic behaviors at four years of age: The UGAAR randomized controlled trial. Environment International 168:107432.
14. Flanagan E, Malmqvist E, Rittner R, et al (2021) Maternal Exposure to Local, Source-specific Ambient Air Pollution During Pregnancy and Autism in Children: a Cohort Study From Southern Sweden. Research Square.
15. Lee KK, Bing R, Kiang J, et al (2020) Adverse health effects associated with household air pollution: a systematic review, meta-analysis, and burden estimation study. The Lancet Global Health 8:e1427–e1434.
16. Luo X-S, Huang W, Shen G, et al (2024) Source differences in the components and cytotoxicity of PM2.5 from automobile exhaust, coal combustion, and biomass burning contributing to urban aerosol toxicity. Atmospheric Chemistry and Physics 24:1345–1360.
17. Gläser E, Kilimann I, Platen M, et al (2025) The economic burden of subjective cognitive decline, mild cognitive impairment and Alzheimer's dementia: excess costs and associated clinical and risk factors. Alzheimer S Research & Therapy 17:.
18. Frech FH, Li G, Juday T, et al (2024) Economic Impact of Progression from Mild Cognitive Impairment to Alzheimer Disease in the United States. The Journal of Prevention of Alzheimer S Disease.
19. Cao L, Zhao Z, Ji C, Xia Y (2020) Association between solid fuel use and cognitive impairment: A cross-sectional and follow-up study in a middle-aged and older Chinese population. Environment International 146:106251.
20. He J, Yang R, Zhong K, et al (2025) Mediation role of healthy lifestyle and socioeconomic status among solid fuels and mild cognitive impairment in middle-aged and older adults: a prospective cohort study. NeuroToxicology 108:338–343.
21. Feeney J, Kenny RA (2024) Use of open fires or closed solid fuel appliances for residential heating and cognitive decline in older adults. Environmental Research 263:120069.
22. Cong X, Zhang J, Sun R, Pu Y (2021) Indoor unclean fuel cessation linked with adult cognitive performance in China. The Science of the Total Environment 775:145518.
23. Chen H, Chen L, Hao G (2021) Sex difference in the association between solid fuel use and cognitive function in rural China. Environmental Research 195:110820.
24. Shen G, Chen Y, Xue C, et al (2015) Pollutant Emissions from Improved Coal- and Wood-Fuelled Cookstoves in Rural Households. Environmental Science & Technology 49:6590–6598.
25. Xu T, Ye X, Lu X, et al (2023) Association between solid cooking fuel and cognitive decline: Three nationwide cohort studies in middle-aged and older population. Environment International 173:107803.

26. Deng Y, Yang T, Gao Q, et al (2021) Cooking with biomass fuels increased the risk for cognitive impairment and cognitive decline among the oldest-old Chinese adults (2011–2018): A prospective cohort study. Environment International 155:106593.
27. Shimada H, Doi T, Lee S, Makizako H (2019) Reversible predictors of reversion from mild cognitive impairment to normal cognition: a 4-year longitudinal study. Alzheimer S Research & Therapy 11:.
28. Yuan J, Tan H, Cheng Y, et al (2023) Air particulate pollution exposure associated with impaired cognition via microbiota gut-brain axis: an evidence from rural elderly female in northwest China. Environmental Science and Pollution Research 31:6398–6410.
29. Tseng T-WJ, Carter E, Yan L, et al (2022) Household air pollution from solid fuel use as a dose-dependent risk factor for cognitive impairment in northern China. Scientific Reports 12:.
30. Karimi L, Mahboub–Ahari A, Jahangiry L, et al (2022) A systematic review and meta-analysis of studies on screening for mild cognitive impairment in primary healthcare. BMC Psychiatry 22:.
31. Naole S, Parikh D, Nayak S, Ramu SP (2025) Evaluating Cognitive Assessment Tools:A comparative analysis of MMSE, RUDAS, SAGE, ADAS and MOCA for early dementia detection
32. Tan M-Y, Wang G-P, Zhu S-X, Jiang L-H (2025) Association between household solid fuel use and cognitive frailty in a middle-aged and older Chinese population. Frontiers in Public Health 13:.
33. He Y, Huang Y, Li R, et al (2024) Switching indoor fuels and the incidence of physical-psychological-cognitive multimorbidity: A prospective cohort study. Ecotoxicology and Environmental Safety 282:116719.
34. Chowdhury SR, Das DC, Sunna TC, et al (2023) Global and regional prevalence of multimorbidity in the adult population in community settings: a systematic review and meta-analysis. EClinicalMedicine 57:101860.
35. Maher BA, O'Sullivan V, Feeney J, et al (2020) Indoor particulate air pollution from open fires and the cognitive function of older people. Environmental Research 192:110298.
36. Liu C, Yang J, Guan L, et al (2024) Intersection of aging and Particulate Matter 2.5 Exposure in real world: Effects on inflammation and endocrine axis activities in rats. International Journal of Endocrinology 2024:.
37. Chen Q, Zhou T, Zhang C, Zhong X (2024) Exploring relevant factors of cognitive impairment in the elderly Chinese population using Lasso regression and Bayesian networks. Heliyon 10:e27069.
38. Mahfoz TMB, Alzahrani M, Shaik RA, et al (2021) Association of indoor air pollution on cognitive dysfunction among elderly. DOAJ (DOAJ: Directory of Open Access Journals) 25:5664–5673.
39. Peng H, Wang M, Wang Y, et al (2024) The association between indoor air pollution from solid fuels and cognitive impairment: a systematic review and meta-analysis. Reviews on Environmental Health.
40. Yang R, He J, Zhong K-K, et al (2024) Relationship of solid fuels use with cognitive function and efficacy of switching to cleaner fuels or using ventilation: A systematic review and meta-analysis. Environmental Research 249:118314.
41. WHO (2025) Dementia
42. WHO (2023) Parkinson Disease
43. Yang W, Hamilton JL, Kopil C, et al (2020) Current and projected future economic burden of Parkinson's disease in the U.S. Npj Parkinson S Disease 6:.

44. Li M, Ma C, Wu C (2023) Association between solid cooking fuel use and dementia in older Chinese adults: the mediating effect of depression. International Journal of Environmental Health Research 34:779–791.
45. Pu F, Hu Y, Li C, et al (2023) Association of solid fuel use with a risk score capturing dementia risk among middle-aged and older adults: A prospective cohort study. Environmental Research 218:115022.
46. Wu H, Lin Z, Zhang C, et al (2025) Biological age acceleration mediates effects of household air pollution from solid fuels on dementia risk: A prospective cohort study in China. Environmental Pollution 381:126603.
47. Oudin A, Segersson D, Adolfsson R, Forsberg B (2018) Association between air pollution from residential wood burning and dementia incidence in a longitudinal study in Northern Sweden. PLoS ONE 13:e0198283.
48. Oudin A, Raza W, Flanagan E, et al (2024) Exposure to source-specific air pollution in residential areas and its association with dementia incidence: a cohort study in Northern Sweden. Scientific Reports 14:.
49. Bernabe-Ortiz A, Carrillo-Larco RM (2024) Estimating the prevalence, factors, and conditions associated with Parkinson disease: a population-based study in Peru. Cadernos De Saúde Pública 40:.
50. WHO (2025) Depressive disorder (depression)
51. WHO (2025) Anxiety disorders
52. Shao, J., Ge, T., Liu, Y., Zhao, Z., & Xia, Y. (2021). Longitudinal associations between household solid fuel use and depression in middle-aged and older Chinese population: A cohort study. Ecotoxicology and Environmental Safety, 209, 111833.
53. Banerjee, M., Siddique, S., Dutta, A., Mukherjee, B., & Ray, M. R. (2012). Cooking with biomass increases the risk of depression in pre-menopausal women in India. Social Science & Medicine, 75(3), 565–572.
54. Liao W, Liu X, Kang N, et al (2022) Effect modification of kitchen ventilation on the associations of solid fuel use and long-duration cooking with the increased prevalence of depressive and anxiety symptoms: The Henan Rural Cohort Study. Indoor Air 32:.
55. Zhang J, Zhang C, Yuchi Y, et al (2025) Associations of solid fuel exposure and kitchen ventilation with mental disorders in rural areas: Evidence from a prospective cohort study. Journal of Affective Disorders 387:119522.
56. Zuñiga RAA, Coca SM, Foláyan MO, et al (2025) Association between biomass fuel use and depression symptoms in the adult population of Oaxaca, Mexico. Diseases 13:47.
57. Diehr PH, Thielke SM, Newman AB, et al (2013) Decline in health for Older Adults: Five-Year change in 13 Key Measures of Standardized Health. The Journals of Gerontology Series A 68:1059–1067.
58. Kawabata E, Tilling K, Groenwold RHH, Hughes RA (2023) Quantitative bias analysis in practice: review of software for regression with unmeasured confounding. BMC Medical Research Methodology 23:.
59. O'Grady C (2025) Low-quality papers surge thanks to public data and AI. Science 388:807–808.
60. Naddaf M (2025) Low-quality papers based on public health data are flooding the scientific literature. Nature.
61. Mitra S, Sagiraju M, Pradhan H, et al (2025) The cognitive toll of household air pollution: cross–sectional associations between polluting cooking fuel use, cognitive functions and brain MRI in a rural aging population from Karnataka, India. The Lancet Regional Health - Southeast Asia 39:100624.
62. Sinton J (2005) Review of China's National Improved Stoves Programs. In: Improved Biomass Cooking Stoves.

63. Li Y, Qiao L, Liu M, et al (2024) Access to affordable and clean domestic heating: A critical review on rural clean heating transformation in China's Jing-Jin-Ji and its surrounding areas. Energy and Buildings 323:114829.
64. Sofi IA, Hassan R, Fazili AI, Qadiri M (2025) wl-7000-Impact of Pradhan Mantri Ujjwala Yojana: Evidence from Jammu and Kashmir. Economic and Political Weekly/Economic & Political Weekly 60:.
65. Adhikari B, Kang SSY, Dahal A, et al (2025) Acceptability of improved cook stoves-a scoping review of the literature. PLOS Global Public Health 5:e0004042.
66. US EPA - Office of Inspector General (2023) The EPA's residential wood heater program does not provide reasonable assurance that heaters are properly tested and certified before reaching consumers
67. Copland S (2025) Dismay as Government refuses to progress wood heater phase out - Conservation Council ACT Region. In: Conservation Council ACT Region.
68. Simões M (2025) New York moves forward in banning gas equipment in New Construction—For now. In: City Limits.
69. Puzzolo E, Fleeman N, Lorenzetti F, et al (2024) Estimated health effects from domestic use of gaseous fuels for cooking and heating in high-income, middle-income, and low-income countries: a systematic review and meta-analyses. The Lancet Respiratory Medicine 12:281–293.

Recycling Library as Fulfillment of the Right to Education and a Healthy Sustainable Environment

Alfin Dwi Novemyanto[1*], *Rismawati* Nur[1], and *Retno* Meilani[1]

[1]Universitas Gadjah Mada, Yogyakarta, Indonesia

Abstract. The fulfillment of the right to education and a sustainable healthy environment still faces serious obstacles in rural areas of Indonesia, one of which is due to limited literacy facilities and waste management, Wonorejo Public Elementary School No. 04 is a concrete example of an elementary school that does not have a library, while the level of household waste production is quite high and has the potential to cause environmental pollution. This study aims to examine the implementation of the Recycling Library as an innovative model to fulfill the right to education while strengthening ecological awareness based on school empowerment. The method used is community-based participatory research (CBPR) with a socialization-monitoring system approach through socialization, assistance, and hybrid monitoring. The results of the study show the formation of a functional library with a collection of 491 donated books, supporting tools from recycled waste, and the revitalization of previously unused school space. This program improves student literacy, fosters a culture of self-help, and reduces the potential for environmental pollution through school-based waste management. The conclusion of this study confirms that the Recycling Library is an instrument for simultaneously fulfilling the right to education and the right to a healthy environment.

1 Introduction

Education is a fundamental right guaranteed in Article 31 of the 1945 Constitution of the Republic of Indonesia [1] and Article 26 of the Universal Declaration of Human Rights (UDHR) [2], which affirms that everyone has the right to education without discrimination. However, in rural areas of Indonesia, this right is still hampered by limited supporting facilities, particularly the absence of libraries as centers of literacy and learning. Based on observations, there are elementary schools that do not yet have library facilities, such as Wonorejo Public Elementary School No. 04 in Karanganyar Regency, Central Java, but they do have unused rooms. This condition hinders the development of 21st-century skills such as critical thinking, creativity, and information literacy as emphasized by UNESCO [3]. UNICEF also emphasizes that limited access to educational facilities widens the gap in learning quality between social groups [4].

* Corresponding author: alfindwinovemyanto@mail.ugm.ac.id

© The Authors, published by EDP Sciences. This is an open access article distributed under the terms of the Creative Commons Attribution License 4.0 (https://creativecommons.org/licenses/by/4.0/).

In addition to educational limitations, environmental problems worsen the quality of life and health of the community around the school. According to observations, Wonorejo Public Elementary School No. 04 is located in an area with a high volume of household waste, but waste management is still minimal, as evidenced by the fact that the final disposal site located in the village exceeds capacity, causing potential pollution and disease. According to the results of bibliometric research, waste management in schools is an important component of sustainable development because education serves as a trigger for awareness and the formation of environmental habits for future communities [5]. A cross-country study in Kenya also shows that school involvement in waste management through a participatory approach influences student participation in educational activities and demonstrates the potential of waste collection as a valuable resource [6]. The collective approach of schools to waste management reflects the principles of the circular economy and sustainable education, while reinforcing the urgency of integrating environmental management into the academic context. Therefore, the school-based approach at Wonorejo Public Elementary School No. 04 can be an effective model in responding to environmental challenges while strengthening the right to education.

The Recycling Library program is an innovative solution to address educational limitations and environmental problems at the local level. This model utilizes recycled inorganic waste to create library support tools, revives used books that are still in good condition, and revitalizes unused school spaces. From a legal perspective in Indonesia, this idea is in line with Article 28H paragraph (1) of the 1945 Constitution of the Republic of Indonesia regarding the right to a good and healthy environment, as well as Article 31 of the 1945 Constitution of the Republic of Indonesia regarding the right to education. Law No. 32 of 2009 concerning Environmental Protection and Management also emphasizes the importance of community participation in environmental management. Through the involvement of teachers, students, and educational staff, this program reflects the principles of self-help and community empowerment. Thus, the Recycling Library not only provides physical facilities but also builds collective capacity based on independence and sustainability.

The synthesis of the urgency of education and the environment shows that the Recycling Library is an instrument for fulfilling the right to education and a healthy, sustainable environment through the empowerment of schools as subjects of development. Schools are no longer positioned as objects of policy but as active actors capable of creating local solutions to structural limitations. The novelty of this program lies in its ability to connect the issues of literacy, the circular economy, and a rights-based approach in a single community-based empowerment model. As theoretical evidence, Zhang and Perkins in their study assert that a community-based empowerment education model enables ecological environmental transformation through active participation and local social capital [7]. By integrating literacy, sustainability, and community participation, the Recycling Library demonstrates a significant academic contribution to the international literature on rights-based education, environmental law, and sustainable development. In this way, schools are empowered to improve their quality, build independence, and strengthen a culture of self-help in responding to global challenges.

2 Research Method

2.1 Design and Methodological Framework

This study uses a community-based participatory research (CBPR) approach that focuses on empowering school subjects as the main actors [8]. This approach is relevant because it places schools, teachers, and students not merely as objects of intervention, but as subjects who actively build solutions to educational limitations and environmental problems. The program was implemented through a socialization-monitoring system model developed by the team, utilizing digital technology (*Google Meet, WhatsApp*, and *digital manuals*) to facilitate interaction, knowledge transfer, and ongoing mentoring. The socialization process uses the guideline book (Figure 1) as an official reference that sets out the rules, procedures, standards, and implementation methods to ensure that the activities are carried out uniformly, systematically, and in accordance with their objectives. This framework is in line with the rights-based approach principle, which emphasizes the fulfillment of the right to education and a healthy environment in a participatory and inclusive manner.

Fig. 1. Recycling Library Manual

2.2 Implementation Stages

The program is implemented through several main stages (Figure 2). The first stage involves socialization accompanied by tutorials on making recycled tools so that partners understand the principles of sustainability and independence. At this stage, the team also presents material on the concept, benefits, and procedures for establishing a Recycling Library, and distributes digital guidebooks as practical references. Next, school partners implement the program based on the socialization materials and guidelines provided. To ensure the sustainability of the implementation, the team provides hybrid assistance and monitoring through *Google Meet, WhatsApp,* and direct visits to the location to ensure that schools are able to run the program independently. This monitoring also serves as a forum for evaluating progress, discussing obstacles, and strengthening the capacity of partners. Through this stage, partner schools have successfully established Recycling Libraries that serve as literacy facilities and media for sustainable environmental education.

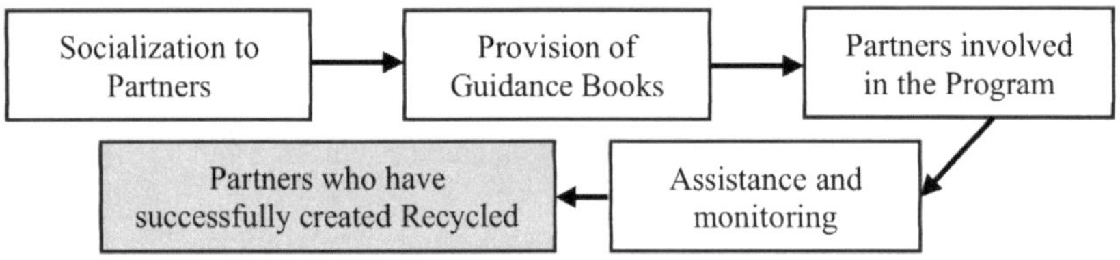

Fig. 2. Implementation Stages

2.3 Location, Subject, and Object of the Program

The program was implemented from August 23 to September 24, 2020, at Wonorejo Public Elementary School No. 04, Wonorejo Village, Gondangrejo District, Karanganyar Regency, Central Java, Indonesia (Figure 3). The subjects of the program were the school community, consisting of teachers, students, and educational staff, while the objects of the program included waste management, the use of used books, and the revitalization of unused school spaces. The intervention focused on the construction of a dual-purpose Recycling Library, which serves as a means of fulfilling the right to education and as an instrument for sustainable environmental management.

Fig. 3. Wonorejo Public Elementary School No. 04

2.4 Data Collection and Analysis Techniques

Data was collected through activity documentation, hybrid interviews, and reflective reports from partners during the implementation process. Documentation included photos, videos, and notes on the process of creating the Recycling Library by the school community. Online interviews were conducted using *Google Meet* and *WhatsApp* (Figure 4) to obtain qualitative data related to the experiences, perceptions, and obstacles faced by partners. Data analysis was conducted using a qualitative descriptive approach through data reduction, presentation, and verification to identify the program's effectiveness. Data validity was strengthened by member checking with partners to ensure that the interpretation was in line with their experiences.

Fig. 4. Communication with partners using digital media

3 Discussion

3.1 Implementation of the Recycling Library as Access to the Right to Education

The implementation of the Recycling Library at Wonorejo Public Elementary School No. 04 began through a mechanism of socialization and assistance based on school community participation. This activity was carried out in a hybrid format, utilizing digital media and face-to-face meetings with a guidebook as a technical guideline for the development of recycling-based literacy facilities. The active involvement of teachers, students, and educational staff in discussions and operational practices reflects an empowerment model consistent with the principles of community education, in which the community is the main actor in local transformation [7]. This participatory approach fosters a sense of community ownership of educational initiatives, thereby creating program sustainability beyond external intervention. The implementation of supporting facilities made from inorganic waste, such as plastic bottles, cardboard, and newspapers, demonstrates effective local innovation in overcoming a lack of resources. Thus, the Recycling Library not only provides learning facilities but also becomes a medium for internalizing the values of independence and sustainability in community-based education.

The transformation of previously unused school space into a functional library represents the fulfillment of the right to concrete and relevant education (Figure 5). This process involves cleaning, simple renovations, and the use of recycled tools, which are collaboratively produced by the school community. The presence of the library actualizes the principles of Article 31 of the 1945 Constitution of the Republic of Indonesia and Article 26 of the Universal Declaration of Human Rights, which affirm education as a fundamental right of every individual. As a theoretical addition, Barrett et al.'s (2019) study in the World Bank report entitled *The Impact of School Infrastructure on Learning: A Synthesis of the Evidence* concludes that the quality of school infrastructure, including representative libraries, has a significant influence on student learning outcomes in various contexts in developing countries [9]. With libraries, students gain equal access to teaching materials, thereby reducing literacy gaps between regions. This condition confirms that the utilization of local resources-such as space and used goods-can bridge structural limitations in the effective provision of educational facilities.

Fig. 5. Room at Wonorejo Public Elementary School 04 before and after becoming a Recycling Library

The collection of books gathered through donations further strengthens the Recycling Library's function as an inclusive and sustainable literacy center (Figure 6). A total of 491 books from various categories, ranging from thematic books to educational magazines, form the initial collection that enriches students' learning resources. This strategy is in line with Law No. 20 of 2003 concerning the National Education System, which emphasizes equal access through the provision of literacy facilities. Findings from Rawlings et al. show that an increase in the number of books and stationery is directly related to an increase in literacy competence in various contexts in developing countries [10]. By integrating book donations into recycling-based libraries, schools have succeeded in bringing about educational innovations that are not only cost-effective, but also inclusive and environmentally friendly.

Fig. 6. Book Donation Poster and Collection of Donated Books

3.2 Recycling Libraries in the Perspective of the Right to a Healthy and Sustainable Environment

The right to a good and healthy environment is a fundamental right guaranteed by Article 28H paragraph (1) of the 1945 Constitution of the Republic of Indonesia and reinforced by Law No. 32 of 2009 concerning Environmental Protection and Management. This legal instrument affirms that every citizen has the right to enjoy a decent environment, including in the context of education, where learning spaces must be clean, safe, and supportive of student growth and development. This concept is in line with the Stockholm Declaration (1972) and the Rio Declaration (1992), which emphasize the importance of integrating human rights with environmental sustainability [11]. The implementation of the Recycling Library at Wonorejo Public Elementary School No. 04 is proof that school-based waste management can be a means of fulfilling these rights. Plastic, bottle, and paper waste that is recycled into library equipment not only reduces pollution but also creates an environmentally friendly learning ecosystem. Thus, this program demonstrates the direct connection between waste management, constitutional rights, and educational sustainability.

The implementation of the Recycling Library shows a real contribution to preventing environmental pollution and strengthening the ecological awareness of the school community. Before this program was implemented, household waste was not managed optimally, posing risks of accidents, disease, and ecosystem degradation. Through mentoring, teachers and students are taught simple recycling techniques to turn waste into bookshelves and other functional tools (Figure 7). According to recent research by Kowasch (2022), integrating circular economy concepts into education, such as cradle-to-cradle and zero waste issues, strengthens students' understanding of ecological sustainability and builds transformative educational capabilities [12]. Field data shows that cardboard, plastic bottles, and old newspapers have been successfully converted into effective literacy support tools. The integration of this practice makes schools not only beneficiaries but also agents of change in reducing waste volume and building sustainability awareness.

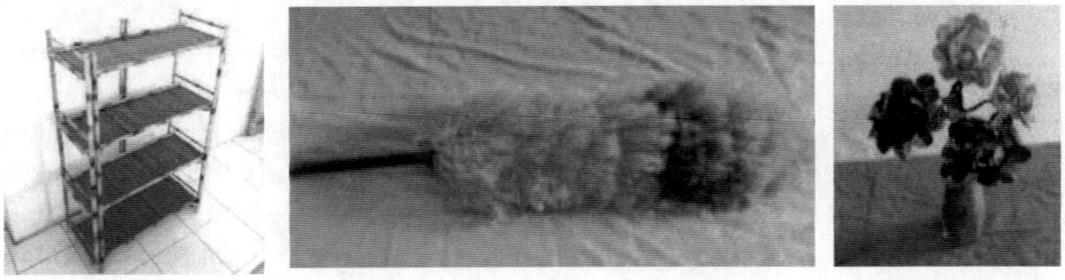

Fig. 7. The results of recycling waste

The Recycling Library contributes significantly to the achievement of the Sustainable Development Goals (SDGs), particularly SDG 4 on quality education and SDG 12 on sustainable consumption and production. By providing a recycling-based library, schools have succeeded in expanding access to literacy without adding to the environmental cost. The book donation, which collected 491 collections, demonstrates the dimension of social participation that strengthens the culture of literacy in schools. According to a study by Zaremohzzabieh et al. (2024), community involvement in waste management through a participatory approach has proven effective in improving the sustainability of waste management practices in developing countries [13]. Within the framework of community-based participatory research (CBPR), the involvement of teachers, students, and educational staff demonstrates the school's capacity as an active subject in maintaining sustainability. Through the integration of literacy, waste management, and community empowerment, the

Recycling Library presents a concrete model for fulfilling the right to a healthy environment as well as sustainable educational transformation.

3.3 Empowering Schools as Subjects of Transformation through the Concept of Self-Help

The active role of teachers, students, and educational staff in the implementation of the Recycling Library shows how schools can become agents of transformation through the concept of self-help (Figure 8). Field results show that school members not only accept the program but are also directly involved in utilizing waste and managing empty spaces to create functional libraries. This approach is in line with the principle of community-based participatory research (CBPR), which emphasizes community involvement in all stages of educational intervention. From an international perspective, a study by Harthun et al. (2019) shows that the involvement of teachers and students in the design and implementation of educational programs increases the relevance and effectiveness of these programs [14]. Through direct involvement, schools build internal capacity to adapt to educational resource constraints. This process confirms a paradigm shift from structural dependence to community-based independence.

Fig. 8. Partners participated in establishing a Recycling Library.

The implementation of the Recycling Library also shows multiple dimensions of empowerment, from waste management to learning tools to strengthening the culture of literacy in schools. Teachers act as facilitators, students as creative implementers, and educational staff as technical supporters, creating a participatory educational ecosystem. According to UNESCO (2025), education for sustainable development must be a core component of school curricula worldwide, with a focus on collaboration and sustainability [15]. Thus, the Recycling Library not only adds to the facilities but also instills the value of sustainability in the entire school community. This empowerment strengthens the function of education as an instrument for shaping environmentally conscious character. This self-help-based approach shows that transformation can be achieved without complete dependence on external intervention. This can be seen in the comparison of the partners' conditions before and after the Recycling Library program (Table 1):

Table 1. Comparison of Partner Conditions Before and After the Recycling Library Program

Aspects Observed	Before the Program	After the Program
Knowledge about waste management	Partners are not yet aware of how to optimize the productive use of waste and used items	Partners understand the concept of utilizing inorganic waste to support the Recycling Library
Environmental awareness	Partners still consider waste to be useless garbage	Partners have ecological awareness that waste can be managed as an alternative resource

Literacy facilities	The school does not have an adequate library as a center for student literacy	The school has a Recycling Library with a collection of 491 donated books and recycled shelves
School community participation	Teachers, students, and educational staff have not been actively involved in managing learning facilities	The school community is directly involved in the creation, maintenance, and development of the Recycling Library
Utilization of school space	There were empty rooms that were unused and had no functional value	The empty space was successfully revitalized into a functional recycling-based library
Program independence	Schools depend on external intervention in the procurement of educational facilities	Schools are able to develop and maintain program sustainability independently based on the principle of self-help
Student literacy access	Students have difficulty accessing reading materials at school	Students gain equal access to teaching materials, thereby encouraging an increase in reading interest and literacy

Furthermore, the empowerment of schools through the Recycling Library demonstrates the close relationship between the right to education and the right to a healthy environment. The use of waste that was previously worthless as a means of supporting the library demonstrates local innovation that is in line with SDG 4 on quality education and SDG 12 on sustainable consumption and production. Within the framework of a rights-based approach, schools are no longer positioned as objects of policy, but rather as subjects capable of producing solutions based on local resources. This shows an important shift from mere access to community control over the fulfillment of fundamental rights. School members not only gain literacy tools, but also build sustainable ecological awareness. Thus, the Recycling Library affirms its novelty in international education literature by connecting education, human rights, and sustainable development through a school empowerment model.

4 Conclusion

The Recycling Library Program at Wonorejo Public Elementary School No. 04 proves that limited educational resources and environmental issues can be overcome through innovation based on school community empowerment. The implementation of the program, which involves teachers, students, and educational staff, demonstrates a paradigm shift from dependence on external intervention to self-help-based independence. The utilization of unused space, donated books, and recycled inorganic waste demonstrates the integration between the fulfillment of the right to education and the right to a healthy environment. The community-based participatory research and rights-based approach make schools active subjects in producing solutions to structural problems. The results of this program are in line with sustainable development goals, particularly SDG 4 on quality education and SDG 12 on sustainable consumption and production. Academically, the Recycling Library affirms its novelty in the literature on education, human rights law, and environmental sustainability. Thus, this program demonstrates the real relevance of literacy, empowerment, and sustainability in the context of basic education.

The authors are grateful to the lembaga Pengelola Dana Pendidikan (LPDP) of the Ministry of Finance of the Republic of Indonesia, which has provided finance support to this publication.

References

1 K. Khairunnisa, D. U. Arini, and T. Anandea, "Human Rights Relationships On Those Children Studied In A School Under Article 31 Of Education Of The Constitution Of The Republic of Indonesia 1945 (at Tunas Mulia Elementary School Gading Serpong)," *Scientific Journal Of Reflection : Economic, Accounting, Management and Business*, vol. 6, no. 1, pp. 56–64, Jan. 2023

2 T. Sharma, "Right To Education As Fundamental Human Right," *International Journal of Information Research and Review*, vol. 04, no. 12, Dec. 2017

3 J. C. González-salamanca, O. L. Agudelo, and J. Salinas, "Key Competences, Education for Sustainable Development and Strategies for the Development of 21st Century Skills. A Systematic Literature Review," *Sustainability 2020, Vol. 12, Page 1036*

4 D. Burde, O. Guven, J. Kelcey, H. Lahmann, and K. Al-Abbadi, *What Works to Promote Children's Educational Access, Quality of Learning, and Wellbeing in Crisis-Affected Contexts*. New York: Department for International Development., 2015.

5 R. Haniva, S. B. Butar, and N. Ambarita, "Waste management in schools as part of sustainable development," *Journal of Sustainability, Society, and Eco-Welfare*, vol. 1, no. 2, Jan. 2024

6 Wambeye K, D. W. Wasike, and P. O. Obino, "Mechanism Of Waste Management In Schools And Its Effect On Student Participation In Education," *International Journal of Development Research*, vol. 12, no. 01, pp. 53491–53495, Jan. 2022

7 Y. Zhang and D. D. Perkins, "Toward an Empowerment Model of Community Education in China," *Adult Education Quarterly*, vol. 73, no. 1, pp. 21–39, Feb. 2022

8 S. S. Coughlin, S. A. Smith, and M. E. Fernandez, *Handbook of Community-based Participatory Research*. New York: Oxford University Press, 2017.

9 P. Barrett, A. Treves, T. Shmis, D. Ambasz, and maria Ustinova, *The Impact of School Infrastructure on Learning: A Synthesis of the Evidence* . Washington: The World Bank, 2019.

10 B. S. Rawlings *et al.*, "Quantifying quality: The impact of measures of school quality on children's academic achievement across diverse societies," *Dev Sci*, vol. 27, no. 5, p. e13434, Sep. 2024

11 E. Akyüz, "The Development of Environmental Human Rights," *International Journal of Environment and Geoinformatics*, vol. 8, no. 2, pp. 218–225, Jun. 2021

12 M. Kowasch, "Circular economy, cradle to cradle and zero waste frameworks in teacher education for sustainability," *International Journal of Sustainability in Higher Education*, vol. 23, no. 6, pp. 1404–1425, Jul. 2022

13 Y. He, Z. Zaremohzzabieh, H. A. Rahman, S. Norkhadijah, S. Ismail, and J. Bin-Qiang, "Applying participatory research in solid waste management: A systematic literature review and evaluation reporting," *Journal of Infrastructure, Policy and Development*, vol. 8, no. 5, p. 5072, 2024

14 M. L. Harthun, P. A. Dustman, L. J. Reeves, F. F. Marsiglia, and M. L. Hecht, "Using Community-based Participatory Research to Adapt keepin' it REAL: Creating a Socially, Developmentally, and Academically Appropriate Prevention Curriculum for 5th Graders," *J Alcohol Drug Educ*, vol. 53, no. 3, p. 12, Dec. 2019

15 UNESCO, "UNESCO declares environmental education must be a core curriculum component by 2025 — Mission 4.7." Accessed: Sep. 09, 2025.

Rethinking Carbon Pricing: Balancing Market Incentives and Environmental Protection Through BUMICX

Retno Meilani*

Universitas Gadjah Mada, Faculty of Law, Yogyakarta, Indonesia

Abstract. This article analyses the paradox of green development in Indonesia, which on the one hand promotes the energy transition agenda and global climate commitments, but on the other hand continues to strengthen its dependence on fossil fuels, particularly coal. Using a mixed approach that combines normative legal analysis and empirical studies of emissions data, fiscal regulations, and energy policies, this research shows that carbon tax instruments in Indonesia still function more as a means of political-economic legitimacy than as effective instruments for controlling emissions. The cap-and-tax model that has been adopted risks turning emissions into a financial commodity rather than an ecological limit that must be adhered to. To address these weaknesses, this article proposes the Balanced Upgraded Model for Integrated Carbon Taxation (BUMICX) as an innovative policy framework. This model positions carbon tax not only as a source of state revenue, but also as a structural mechanism to encourage green investment, strengthen institutions, and ensure ecological and social justice. The findings of this study confirm that without consistent and certain regulatory reform, market orientation tends to override environmental protection, putting decarbonization efforts at risk of being trapped in rhetoric without real transformation.

1 Introduction

Amidst an increasingly alarming global climate crisis, the greatest irony of the energy transition comes from developing countries that still depend on fossil fuels such as coal for their economic growth, even though it is known to be the most polluting fuel [1], [2]. This situation is exacerbated by the policy of accelerating economic recovery after the COVID-19 pandemic, which paradoxically increases the use of coal in the national energy mix instead of leading to reduced emissions and clean energy diversification. This is marked by the operation of several new coal-fired power plants and the expansion of the nickel industry, which the government has positioned as a national strategic priority [3]. Nickel down streaming is projected to be key for Indonesia to strengthen its role in the global supply chain, especially for electric vehicle batteries and environmentally friendly

* Corresponding author: retnomeilani@mail.ugm.ac.id

© The Authors, published by EDP Sciences. This is an open access article distributed under the terms of the Creative Commons Attribution License 4.0 (https://creativecommons.org/licenses/by/4.0/).

technologies [4]. However, the expansion of this industry relies on coal energy supplies, which creates a paradox. Development that is claimed to be environmentally friendly at the global level is essentially supported behind the scenes by black smoke, making "environmentally friendly" nothing more than a slogan.

Currently, Indonesia ranks as the third largest coal producer in the world [4]. Projections from the Institute for Essential Services Reform (IESR) estimate that coal consumption in Indonesia will continue to increase until 2029, in line with plans to build and operate new coal-fired power plants [5]. This situation has serious implications not only for the domestic energy transition agenda, but also for the achievement of global climate targets [6]. High dependence on coal as the main energy source will pose a major dilemma in the context of sustainable development goals (SDGs).

In environmental law and energy governance perspective, this dilemma highlights a regulatory contradiction. On the one hand, the government has committed to reducing emissions through the Paris Agreement and Nationally Determined Contribution (NDC). On the other hand, domestic regulations still allow ample room for the expansion of fossil fuels. Empirically, coal produces much higher carbon dioxide (CO_2) emissions than oil or natural gas [7]. This fact confirms that coal is not only the "most polluting" fuel, but also the largest source of emissions that hinders the achievement of emission reduction targets, especially greenhouse gases (GHG). Furthermore, based on data from 2019–2023, Indonesia's GHG emissions continue to show an upward trend, reaching 1,053,476 $GgCO_2e$ in 2023, with CO_2 as the largest contributor at 75–76% of total emissions. This fact shows that decarbonization of the energy sector has not been effective, even though the government has formulated a *Net Zero Emission 2060* target. The dominance of CO_2 indicates a high dependence on coal, oil, and gas, meaning that the expected energy transition still faces structural obstacles, namely high industrial energy demand, slow adoption of renewable energy, and a lack of legal incentives to promote carbon efficiency.

Furthermore, the attention of stakeholders, both national and international, to the issue of carbon emissions should not be interpreted merely as a moral or political commitment but must be translated into operational and certain legal instruments. Without a clear and consistent regulatory framework, this attention has the potential to become normative jargon without any enforcement power. For example, the absence of technical norms in the implementation of carbon tax in the Taxation Harmonization Law (UU HPP) shows a discrepancy between domestic regulations and international commitments, which ultimately creates a *gap* in emissions control. The issue of climate change triggered by greenhouse gas (GHG) emissions has become a global agenda. The figures presented are not merely statistics, but a reflection of Indonesia's ambivalent position: on the one hand, as a developing country that still relies on fossil fuels to drive economic growth, and on the other hand, as a signatory to the Paris Agreement with a commitment to significantly reduce emissions [8].

Ultimately, the question arises: are existing legal instruments (particularly carbon taxes) truly designed to reduce emissions, or do they merely serve as a means of political and economic legitimization that maintains the dominance of the energy market? Furthermore, is it true that market incentives are often the top priority, while environmental protection takes a back seat? This study departs from the dilemma faced by developing countries, particularly Indonesia, which is often faced with a choice between economic development and ecological sustainability, as seen in down streaming programs that damage ecosystems and the living spaces of indigenous peoples. By unravelling this paradox, this study does not stop at analysing the problems that occur but also offers the Balanced Upgraded Model for Integrated Carbon Taxation (BUMICX) as an innovative policy framework. BUMICX positions carbon tax not only as a source of state revenue, but also as a structural catalyst that links legal reform, institutional strengthening, and environment-based investment.

2 Methodology

This study uses a mixed approach that places normative legal research as the main basis but is enriched with empirical analysis. In the normative dimension, the study focused on analysing national legislation, international legal instruments, and institutional frameworks relevant to carbon taxation and energy governance. Meanwhile, on the empirical dimension, researchers processed statistical data on carbon emissions, fiscal revenue, and energy transition indicators sourced from government agency reports, international organizations, and academic publications. Data was obtained through directed literature review using a purposive documentary sampling strategy, which involves selecting sources that are substantively relevant to the issue being studied [9]. The analysis was conducted by combining critical thematic discourse and descriptive comparative methods [10]. The integration of these two analyses resulted in a conceptual synthesis that formed the basis for the formulation of the Balanced Upgraded Model for Integrated Carbon Taxation (BUMICX).

3 Discussion

3.1 Effectiveness of Carbon Tax Instruments in Controlling Emissions

Carbon emissions occur due to the high use of coal and fuel oil. This condition causes air pollution that can disrupt human activities. In cases like this, the marginal social costs borne by the community are much greater than the marginal private costs borne by companies. Upon closer examination, the distribution of GHG emissions in Indonesia is dominated by the production/business sector at 92.12%, while households only contribute around 7.88% [4]. This data is legally significant because it reveals the map of legal responsibility, as emissions control regulations should focus on the industrial, energy, and transportation sectors, rather than solely on household consumption behaviour. In other words, if carbon control policies are only imposed through general fiscal instruments without distinguishing the scale of emissions contributions, then regulatory injustice will arise [11], where the legal burden is felt more heavily by parties with smaller emissions contributions. Therefore, to minimize these negative impacts, the government has established a *carbon pricing* mechanism, which can be implemented through an Emission Trading System (ETS) or carbon tax.

The ETS system focuses on the cap-and-trade principle. Meanwhile, carbon tax focuses on setting a price tariff on excess CO2 production. Both alternatives, ETS and carbon tax, have the same goal of reducing emission levels. However, carbon taxes are clearer in terms of administration, such as determining who is liable to pay the tax, the basis for imposing the tax, the tax rate, and the procedures for payment and reporting, compared to the ETS system, which is relatively more complex because it involves the allocation of emission quotas, the trading of quotas between businesses, and market mechanisms that can fluctuate. In addition, the effectiveness of ETS also depends heavily on the setting of appropriate emission limits and market participant compliance, so its management requires more intensive monitoring and supervision than a regular carbon tax.

In Indonesia, the implementation of carbon pricing adopts a cap and tax scheme, which combines the principles of cap and trade with carbon tax. Under this scheme, companies are given a maximum emission limit (cap), and if actual emissions are below this limit, the difference can be used as an asset or traded. Conversely, if emissions exceed the limit, companies are required to pay carbon tax on the excess. However, this pattern shows that this scheme has the potential to undermine environmental protection goals. By giving companies, the option to pay taxes on excess emissions, this system implicitly treats emission limits as a "price that can be bought," rather than as ecological limitations that must be complied with. As a result, large companies with financial capacity can continue to exceed the cap without promoting real emissions reductions. In fact, in the context of Indonesia, which still relies on fossil fuel expansion (coal and new coal-fired power plants), cap and tax risks becoming a "carbon permit to pollute," where tax revenues serve more as financial compensation than as an instrument of structural transformation. Thus, this issue confirms that cap-and-tax is not adequate for developing countries such as Indonesia.

The problems faced by developing countries, particularly Indonesia, cannot be separated from the structural dilemmas at the center of the triangle between the economy, politics, and the environment. These dilemmas are even more apparent in the context of post-COVID-19 economic recovery. Economic pressures often dominate energy policy decisions. In practice, the energy agenda is often dominated by economic considerations, so that legal and fiscal policies are not always in line with environmental and social objectives [12]. This raises the risk that regulatory interventions will be merely symbolic or insufficient in promoting a clean energy transition. The data in the graph shows a significant upward trend in coal supply in Indonesia from 2000 to the projected year of 2025. This increase not only reflects a growing dependence on high-carbon fossil fuels but also contributes to widespread environmental damage caused by downstream projects, particularly in the extractive industry sector. This graph highlights the paradox in the implementation of energy policy in Indonesia, where strategic industrialization efforts are deepening the carbon footprint while worsening ecosystem degradation.

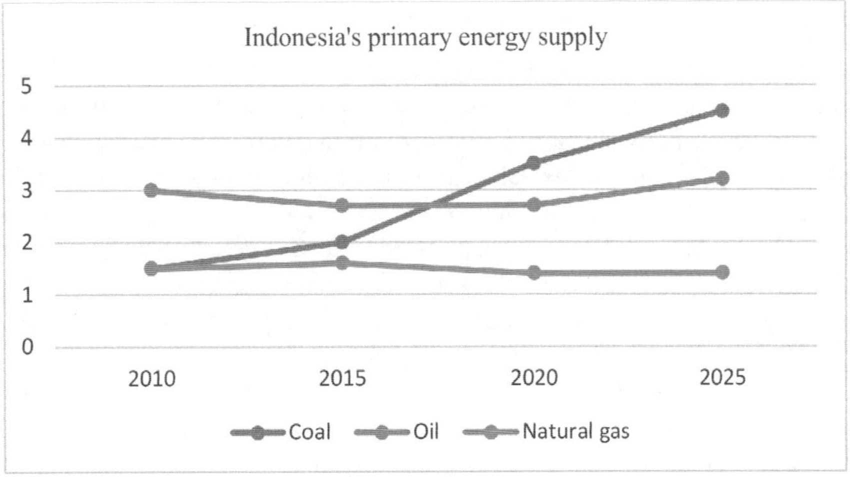

Fig. 1. Indonesia's primary energy supply by source

The graph shows the dynamics of Indonesia's primary energy supply for the period 2000–2025, which consists of three main components, namely coal, oil, and natural gas. The data shows that coal supply has experienced the most significant increase, especially after 2015, even surging sharply in the period 2020–2022, which confirms Indonesia's increasing dependence on fossil fuels with high carbon intensity and high emissions. In contrast, petroleum supply tends to be stable with moderate fluctuations in the range of 2.5–3 EJ, while natural gas has only shown slow growth and relative stagnation after 2015. This pattern reflects the limitations of clean energy diversification. As shown in Figure 2, the map of environmental damage caused by mining activities in Indonesia shows the spatial distribution of areas directly affected by coal, nickel, and gold extraction. Overall, nearly 80% of Indonesia's territory is recorded as having mining activities. Most of these areas show signs of significant environmental damage. In the province of Kalimantan, the region with the highest concentration of coal mines in Indonesia, areas such as Hulu Sungai, Tapin, and Kutai Kartanegara show red zones representing high levels of land damage.

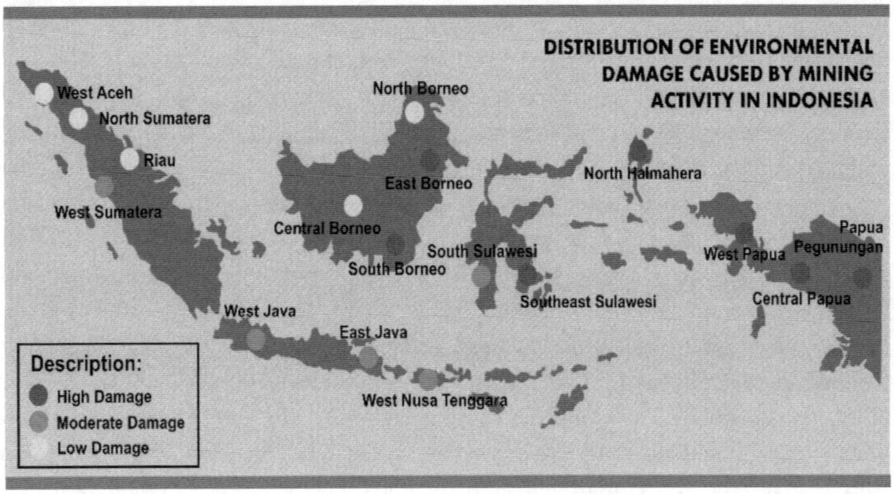

Fig. 2. Distribution of environmental damage caused by mining activity in Indonesia

In Papua, West Papua, and Papua Pegunungan, the map shows two main layers of problems. In Mimika, the Grasberg gold mine is associated with the disposal of tailings into the river system, causing heavy metal pollution and soil degradation, while on Gag Island (Raja Ampat), nickel exploitation has a direct impact on coastal and marine ecosystems. In addition to the loss of marine biodiversity, these activities have implications for increased blue carbon loss due to the destruction of mangrove and seagrass ecosystems that previously functioned as natural carbon sinks. Meanwhile, in Sulawesi, namely South Sulawesi, Southeast Sulawesi, and North Maluku, the map shows the destruction of tropical forests due to the expansion of large-scale nickel mining. The regions of Halmahera, North Konawe, and East Luwu have become environmental degradation hotspots marked by deforestation, declining biodiversity, and coastal sedimentation. These activities not only reduce the carbon absorption capacity of tropical forests but also generate additional carbon emissions from peatland degradation and the use of fossil fuels in nickel processing.

In West Sumatra, West Java, East Java, and West Nusa Tenggara, environmental damage related to gold and coal mining is evident in the form of heavy metal pollution (mercury and cyanide), unreclaim mining pits, and the degradation of old mining sites such as Sawahlunto. In addition to causing soil and water pollution, unrehabilitated land degradation also becomes a source of long-term carbon emissions due to the loss of natural vegetation that functions as a carbon sink.

In general, this map confirms the spatial correlation between mining production concentration and the intensity of environmental degradation. Areas with dominant extractive activities are marked with intense colours on the map, which not only illustrate the level of ecosystem damage, but also represent areas with significant contributions to national carbon emissions. This visualization reinforces the argument that the exploitation of natural resources in Indonesia should not be viewed solely as an economic issue, but also as a global environmental issue, particularly in relation to Indonesia's commitment to reducing carbon emissions and achieving the Net Zero Emission (NZE) target by 2060.

Within the capitalist economic framework, companies and investors are motivated by incentives to maximize profits and minimize costs [2]. Environmental protection, including investment in low-carbon technology or mine reclamation, is often seen as an additional burden that cuts into profit margins. This explains why governments, in pursuit of rapid economic growth, job creation, and increased state revenues, often support the expansion of extractive industries despite their significant environmental impact. One of the most obvious examples is the expansion of the coal and nickel mining industries in sensitive ecosystems such as Raja Ampat, which ironically worsens environmental degradation while reducing the ecosystem's capacity to absorb carbon.

In many cases, environmental regulations are seen as obstacles to economic growth [13]. In Indonesia, this mindset finds its manifestation in concrete policies such as the Omnibus Law on Job Creation. This regulation was designed to simplify bureaucracy, including environmental licensing, with the pretext of attracting investment and encouraging job creation. However, when analysed critically, the Omnibus Law becomes clear evidence of how market incentives are systematically placed above environmental protection. Ultimately, the Omnibus Law reinforces a cycle in which profit always trumps sustainability.

3.2 The BUMICX Model as an Innovative Environmental Control Policy

Given The complexity and challenges in implementing carbon reduction in Indonesia, there is a clear need for a more integrated, adaptive, and solution-oriented policy framework [14]. In this context, this article proposes a Balanced Upgraded Model for Integrated Carbon Taxation (BUMICX) as a holistic framework for developing countries such as Indonesia, which face a dilemma between economic growth, emissions reduction, and industrial sector readiness.

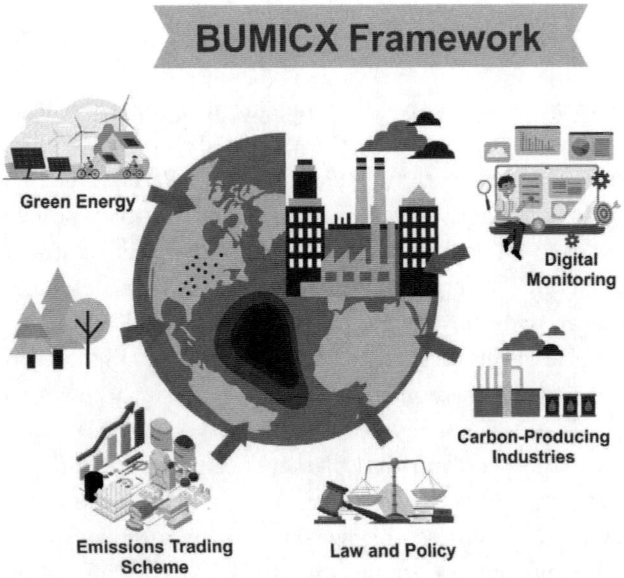

Fig. 3. The BUMICX framework

Figure 3 shows The BUMICX model integrates carbon financing with the Emissions Trading Scheme (ETS) system. However, the ETS system adopted by BUMICX differs from previous ETS systems. In this ETS trading system, companies that exceed their carbon emission limits are no longer required to purchase quotas but instead invest directly in projects that promote green transition. This mechanism transforms pollution fines into a strategic source of financing. For example, a factory that emits 100 tons of CO2 above its threshold is required to deposit an amount equivalent to the value of those 100 tons of carbon. These funds do not disappear as fines but are channelled into a designated green investment fund. The funds are then proactively used to finance the construction of solar, wind, or biomass power plants, fund research and development of low-carbon technologies, or support the modernization of factory machinery to make it more energy efficient.

This approach fundamentally changes the company's mindset from simply "paying to pollute" to "financing solutions," making every excess emission a tangible contribution to creating a more sustainable future. This system can also mitigate excessive carbon use. Another strength of BUMICX is the accurate and transparent integration of digital-based monitoring and enforcement systems, which minimizes the risk of data manipulation or tax evasion [15]. The implementation of BUMICX can be carried out in stages to manage economic and social risks. This phased approach allows industry and society to adapt gradually, while giving the government room to evaluate the impact of the policy and make any necessary adjustments. Overall, BUMICX is an innovative and comprehensive approach that addresses the limitations of conventional carbon taxes in developing countries. Through a combination of tariffs, hybrid ETS mechanisms, social compensation, strategic revenue allocation, digital monitoring systems, and phased implementation, this model can balance emissions reduction, economic growth, and social equity.

4 Conclusion

This study shows that the green growth paradigm in Indonesia still faces fundamental contradictions between the economic growth agenda and environmental protection commitments. Although the government has ratified the Paris Agreement and set a target of Net Zero Emissions by 2060, the implementation of energy policies has shown an increase in dependence on coal, especially to support the downstreaming of strategic industries such as nickel. The legal instruments that have been designed, particularly the carbon tax with a cap-and-tax scheme, have not functioned optimally as an emissions control mechanism. However, these instruments place more emphasis on the fiscal aspect as a source of state revenue than on their role in significantly reducing greenhouse gas emissions.

The results of the study confirm that market incentives are still placed higher than ecological sustainability, so that decarbonization policies have the potential to remain at the level of rhetoric. To address this weakness, this article proposes the Balanced Upgraded Model for Integrated Carbon Taxation (BUMICX), an innovative framework that combines carbon pricing with green investment obligations, solution-based emission trading mechanisms, and a transparent digital monitoring system. This model is expected to shift the paradigm from simply “paying for pollution” to “financing solutions,” so that every excess emission can contribute significantly to low-carbon development. Therefore, the success of the energy transition in Indonesia can only be achieved if there are consistent legal reforms, strict law enforcement, and political commitment to place environmental interests on par with economic and social interests.

The authors are grateful to the Lembaga Pengelola Dana Pendidikan (LPDP) of the Ministry of Finance of the Republic of Indonesia, which has provided financial support to this publication.

References

1. M. Khastar, A. Aslani, and M. Nejati, "How does carbon tax affect social welfare and emission reduction in Finland?" Energy Reports, **6** (2020)
2. N. Bellanca, "Capitalism," in Handbook on the Economics of Reciprocity and Social Enterprise, (Edward Elgar Publishing Ltd., 2013)
3. Transparancy International, "Industri Keruk Nikel : Korupsi Struktural dan Dampak Multi Dimensinya," 2024
4. BPS RI, Indonesia Energy Flows Accounts and Greenhouse Gas Emissions Accounts, **5** (2025)
5. H. N. Jong, "Indonesia's coal burning hits record high and 'green' nickel is largely why?" 2023
6. R. J. Heffron, D. McCauley, and B. K. Sovacool, "Resolving society's energy trilemma through the Energy Justice Metric," *Energy Policy,* **87** (2015)
7. L. Sugiharti, R. Purwono, M. R. Primanthi, and M. A. Esquivias, "Indonesia industrial productivity growth: Evidence of Re-industrialization or De-industrialization?" *Periodica Polytechnica Social and Management Sciences*, **27** (2019)
8. N. M. Purnama, C. Sugiarto, R. N. Aini, and M. Putri, "Evaluasi Kesiapan Dan Tantangan Implementasi Pajak Karbon di Indonesia Dalam Perspektif Keadilan Iklim dan Pembangunan Berkelanjutan," *Jurnal Penelitian Nusantara*, **1** (2025)
9. T. Palys, "Purposive Sampling," *The Sage Encyclopedia of Qualitative Research Methods*, 2 (2008)
10. B. Lawless and Y. W. Chen, "Developing a Method of Critical Thematic Analysis for Qualitative Communication Inquiry," *Howard Journal of Communications*, 30 (2019)
11. N. Anyachebelu and P. O. Nweke, "Challenges and prospects of implementing international environmental treaties in developing nations amid economic and political constraints," *Article in Journal of Critical Reviews*, (2020)
12. J. Pan, J. L. Cross, X. Zou, and B. Zhang, "To tax or to trade? A global review of carbon emissions reduction strategies," (2024)
13. A. M. Tavares and Syahrani, "Nasib Target Emisi Indonesia: Pelemahan Instrumen Lingkungan Hidup di Era Pemulihan Ekonomi Nasional," *Jurnal Hukum Lingkungan Indonesia*, **7** (2020)
14. H. Maghfirani, N. Hanum, and D. Amani, "Analisis Tantangan Penerapan Pajak Karbon Di Indonesia," *Juremi: Jurnal Riset Ekonomi*, **1** (2022)
15. Y. H. Yusoff, I. Nadilah, M. K. Anwar, R. A. Yustiansyah, R. H. Utama, and M. Dahlan, "Optimizing the Implementation of Carbon Tax in Reducing the Impact of Environmental Pollution," *Accounting and Finance Research*, **13** (2024)

IoT-based Multimodal Borewell Monitoring

Tanmay Bhatt[1*], *Jagan M.* Reddy[1], *Tarun* Pragada[1], *Sachin* Chaudhari[1], and *Kalpana* Ramesh[2]

[1]International Institute of Information Technology, Hyderabad, India
[2]The Rainwater Project

Abstract. Excessive groundwater extraction has led to a significant decline in water tables, highlighting the need for continuous monitoring. This study presents an Internet of Things (IoT)-based approach for real-time estimation of groundwater levels in borewells. In addition to water level measurements, parameters such as motor current and rainfall are tracked to evaluate their influence on groundwater dynamics. To assess practical feasibility, five monitoring nodes were deployed across a small educational campus in Hyderabad, India. Data collected over a two-month period indicate that the proposed low-cost system is both reliable and effective in real-world conditions, while also revealing interesting correlations between groundwater level variations and the monitored parameters.

1 Introduction

Groundwater constitutes a vital source of freshwater, particularly in regions where surface water availability is limited or unreliable. In countries like India, where significant spatial and temporal variability in rainfall exists, groundwater extracted through borewells serves as a critical lifeline for both domestic and agricultural needs [1, 2]. The widespread dependence on borewells has grown substantially over the past few decades, largely due to their ability to provide decentralised access to water in areas beyond the reach of piped or surface water infrastructure. Given this dependency, the systematic monitoring and management of borewell usage has become crucial. Continuous data collection regarding borewell performance, water table levels, and patterns of extraction is essential to safeguard these underground resources from depletion. Overexploitation without appropriate regulation has already led to falling groundwater tables in several regions, rendering numerous borewells dry and non-functional
[3].

Traditional field methods for measuring groundwater levels—such as graduated measuring tapes and manual well sounders—remain popular because they are inexpensive and straightforward; however, these approaches are inherently labour-intensive, vulnerable to operator error, and poorly suited to long-term deployments that demand high sampling frequency or large spatial coverage [4]. Recent studies emphasise that the prevailing spatiotemporal resolution of groundwater data in India is not enough for capturing rapid drawdown events and seasonal recharge pulses. Hourly to sub-hourly observations obtained

* Corresponding author: tanmay.b@research.iiit.ac.in

© The Authors, published by EDP Sciences. This is an open access article distributed under the terms of the Creative Commons Attribution License 4.0 (https://creativecommons.org/licenses/by/4.0/).

from autonomous Internet-of-Things (IoT) nodes can fill this gap, yielding insights that periodic manual measurements can't provide [5]. Nevertheless, the global literature on low-cost, opensource IoT solutions for groundwater monitoring remains relatively thin, and most existing prototypes address only a subset of hydro-physical variables or are demonstrated at limited scales [6–10].

Espinoza et al. developed an economical real-time system that measures piezometric depth and temperature while simultaneously logging barometric pressure to enable accurate head correction [6]. Chan et al. proposed two Arduino-based designs that contrast a differential-pressure approach against a single absolute-pressure sensor, thereby illustrating cost–accuracy trade-offs for small community installations [7]. Calderwood et al. showed how multiple pressure-sensor nodes can be networked to form a distributed observatory capable of on-board data processing and local fault detection [8]. Complementing these efforts, Kombo et al. demonstrated that long-range, low-power wide-area networking (LoRaWAN) can reliably relay water-level data from remote boreholes to cloud servers, a feature crucial for regions lacking cellular connectivity [9]. Finally, Tanmay et al. introduced a novel string tension mechanism tailored for borewell shafts, enabling continuous depth tracking without submerging electronic components [10].

Most existing groundwater level monitoring solutions [6–9] rely on pressure sensors, which present several limitations [11]. These sensors operate within a fixed pressure range and must be calibrated according to expected water level variations. They are prone to issues such as drift, hysteresis, and temperature sensitivity, requiring regular calibration and correction. Prolonged exposure to water can also lead to material degradation or failure due to chemical reactions. Additionally, many of these systems do not incorporate complementary parameters like motor usage or rainfall, which are essential for a more holistic understanding of borewell dynamics.

This paper extends the work in [10] by using the data collected using these devices in borewells in the IIIT Hyderabad campus. The major contributions of this paper are • The IoT-enabled borewell water level monitoring system presented in [10] was specifically redesigned for extraction borewells. Five such devices were deployed across the IIIT Hyderabad campus, India, and operated over a two-month period.

- Complementary IoT-based current sensing modules were installed on the motors responsible for groundwater extraction. These enabled the monitoring of motor activity, providing insights into water pumping patterns and usage behaviour.
- To assess natural recharge, an IoT-based rain gauge was also developed and installed on campus to record local precipitation levels.
- The collected data was analysed in conjunction with the site's topographical contours, allowing inferences to be drawn regarding the spatial and temporal variations observed in water levels and motor activity.
- The relationship between variables like water depth, precipitation, terrain and motor current was explored to determine correlated variables.

The proposed IoT-based approach streamlines the process of trend analysis of water consumption by eliminating the need for specialised equipment, manual labour, and significant time investments traditionally required to monitor these vital parameters.

The structure of the paper is as follows: Section II contains the hardware overview of the device. It is followed by an explanation of the data collection and processing in Section III. The results are discussed in Section IV. Section V concludes the paper and delves into the future directions of this work.

 t p s/dl o.gr5d 0g 01(e3sconf606)) 900100(

2 Hardware Overview

2.1 Water Level Device

The device uses a floating bob suspended in the borewell pipe by virtue of a nylon thread. The bob is lowered using a JGY-370 12V geared DC motor. Fig. 1b shows the high-level block diagram of a node. The custom PCB consists of an ESP32 WROOM microcontroller, a DRV8231ADSGR H-bridge motor driver for controlling the DC motor, a DRV5033AJQLPGM Hall effect sensor for counting magnetic pulses and a Quectel UC200 RF transceiver module, which establishes GPRS connectivity via a 2G/3G cellular network to push data to a cloud-based server on *ThingSpeak*[12]. In addition to these major components, the PCB consists of additional circuitry such as resistors, capacitors, and voltage regulators to ensure the proper functioning of these parts. The device is powered using a 12V, 3A power supply. The specifications of the device components are listed in Table 1.

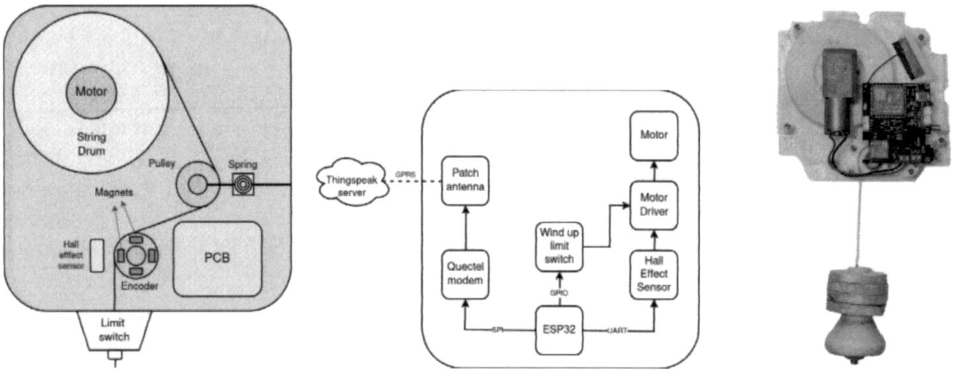

(a) Schematic diagram (b) Block diagram (c) Assembled node

Fig1. Schematic diagram, block diagram, and assembled borewell node.

Fig. 1a shows the schematic diagram of the device. The string drum is attached to the motor, and it turns based on whether the bob needs to be lowered down or pulled up. The string then passes through a pulley connected to a spring, which keeps the string in tension. Ultimately, the string passes through a rotary encoder, which has 4 tiny magnets embedded on the circumference. When this encoder makes 1 complete revolution, the hall effect sensor picks up 4 individual pulses due to the change in magnetic flux. These pulses are used to measure the length of the string, which is being lowered or pulled, thus giving us the distance as

$$distance = \frac{\pi N_{pulses} D_{pulley}}{4}, \qquad (1)$$

where N_{pulses} are the total number of pulses recorded in a reading and D_{pulley} is the diameter of the pulley, which is 29 mm in this case. Fig. 1c shows an actual assembled node.

2.2 Secondary Devices

A DFRobot 'Gravity: tipping bucket rainfall sensor' rain gauge was installed [13]. It is a tipping bucket style rain gauge of size 118 mm ×59 mm ×80 mm, which gives a reading of t pp s/dl o.gor5d 0g 01(de3sconfd606)) 900100(

Table 1: Component specifications

JGY-370 DC Motor	
Rated voltage	12 V
Load speed	128 rpm
Load current	180 mA
Load torque	0.55 kg.cm
Stall torque	2.22 kg.cm
Stall current	1 A
Ratio	37.3

DRV8231ADSGR Motor driver	
Supply voltage	4.5-33 V
Output voltage	0.7-35.7 V
Operating current	3 mA
Output current	3.7 A

DRV5033AJQLPGM Hall effect sensor	
Supply voltage	2.5-38 V
Output voltage	0-38 V
Output current	0-30 mA

resolution 0.28 mm. The operating range of the sensor is 3.3-5 V. It is connected to an ESP32 microcontroller, which polls the rain gauge every 2 minutes and sends the rainfall data in mm via a SIM800L GSM module.

A YHDC SCT-013-000 CT current sensor [14] monitors the motors that pump water from the stepwells. The sensor is interfaced with an ESP32 microcontroller, which sends the motor on/off status with the help of the SIM800L GSM module every 10 minutes. The active duration of motor pumping is tracked using these current sensors. The sensor is rated for 100 A RMS, 120 A peak-to-peak input and converts it to a 50 mA output, with a turn ratio of 1:2000.

3 Data Collection and Processing

Five nodes were deployed in the IIIT Hyderabad campus at locations shown in Fig. 2. Three of these nodes, *A, B and C*, are in extraction borewells where there are motors present for pumping the water. The remaining two, *D and E*, are in injection borewells without motors to study the interaction of these with surrounding extraction borewells. The nodes take readings every 15 minutes. The devices send raw data to the ThingSpeak server, which needs to be processed before inferences are obtained.

3.1 De-trending the data

When a node is deployed in an extraction borewell, the bob cannot be pulled up to the reset position each time since it can get tangled with the pipe which is being used for pumping water. So, in order to get readings, the bob is allowed to reach the water level, the distance is recorded, and then the bob is pulled up by a fixed distance of 2 meters. This ensures that while the device is sleeping for the 15-minute interval, the bob remains suspended and the string remains in tension. This fixed distance was decided based on the collected data, where the water level does not rise as drastically as 2 meters in 15 minutes. Doing this means that in each reading, this fixed amount of 2 meters gets added as an overhead, which causes an increasing trend to develop in the data being sent to ThingSpeak. This trend is removed by subtracting 2 meters cumulatively from the data to get the original water level readings.

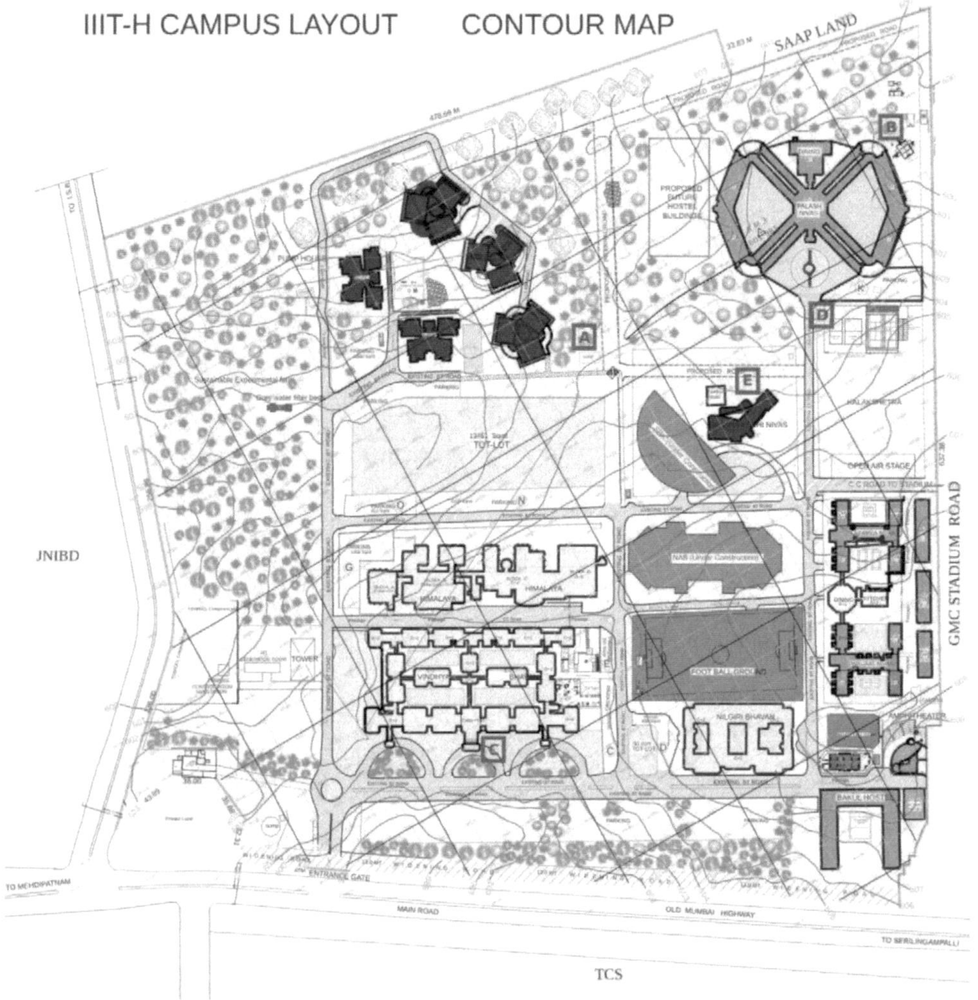

Fig. 2. Deployment map with campus contours

(a) Borewell A (b) Borewell B (c) Current sensor (d) Rain gauge

Fig. 3. Device deployment

3.2 Data smoothening

The raw data from ThingSpeak is processed to smooth it and remove outliers, which might be generated due to device malfunctions. Since we are working with hydrological time series data, a windowed median approach is taken to remove outliers and smooth the data[15]. The window size for borewells A and B is 5, 7 for borewells C and D and 3 for borewell E. The window sizes were determined based on the nature of the data via the trial-and-error method.

3.3 Rain data resampling and thresholding

The rain data, which is being collected every 2 minutes, was resampled to an hourly frequency by cumulatively adding the readings collected over one hour. This was done to get a realistic sense of rainfall. Moreover, these hourly readings were thresholded with the threshold being 2 mm of rain. Only readings greater than this would be considered as a legitimate precipitation event. This ensures that false positives which may have been registered are removed.

3.4 Calculating water table

The readings returned by the devices are with respect to the ground level at that point. This may not provide the complete picture to perform a comparative study; hence, the altitudes of the geographical locations of these borewells are also taken into account. Fig. 3 shows the contours and the elevation of the deployment locations with respect to the mean sea level. *A* is at 603.71 m, *B* is 599 m, *C* is 601.35 m, *D* is 601.55 m and *E* is at 603.74 m.

4 Results

4.1 Water level vs topology

4.1.1 Extraction borewells

The extraction borewells on campus, namely borewells A, B and C, act as the primary sources of groundwater. The plots in Fig. 4a-4c shows the water depth trends in these borewells. It can be observed that the duration for which the motor is on shows a rise in the water depth, meaning a decrease in the water level. When the motor is turned off, the level recuperates back, which is characterised by a fall in the depth. The red dashed line depicts the rainfall in mm.

Seeing the trend in borewell A from Fig. 4a, it shows an increase in the water depth in the month of April, due to the absence of rainfall and excessive pumping owing to the water shortage in the municipal water throughout the city of Hyderabad. The noteworthy observation comes in the month of May when there were rains due to pre-monsoon setting in but the borewell still shows an increasing trend in water depth, which suggests that the aquifer which is connected to this particular borewell is likely to be a deep, isolated one which is not being recharged by the shallow aquifer which gets recharged by surface runoff and rains. This can be attributed to the borewell lying on a ridge point, according to Fig. 2, meaning the water naturally tends to flow away from this point. Moreover, it is likely that a rocky layer separates this deep aquifer from the shallow aquifer.

The trend in borewell B also follows a similar increasing trend in Fig. 4b, as was seen with borewell A in the month of April. This borewell has a lower depth compared to borewell A, which can be attributed to its lower geographical altitude, as can be observed from the contours in Fig. 2. This can imply the possibility that the water from the surrounding high lying areas flows towards this natural low point, which in turn recharges the aquifer connected to this borewell, thus resulting in a lower depth at which water is found. Upon the setting in of pre-monsoon in May, we observe a plateauing of the depth even though the pumping continues at the same rate, which hints towards a substantial recharge of the underground aquifer from the shallow aquifer. The contour lines also suggest that the surface runoff from the surrounding areas is likely to collect near this borewell, which further leads to better incorporation into the shallow aquifer at this location and, in turn, the deeper aquifer.

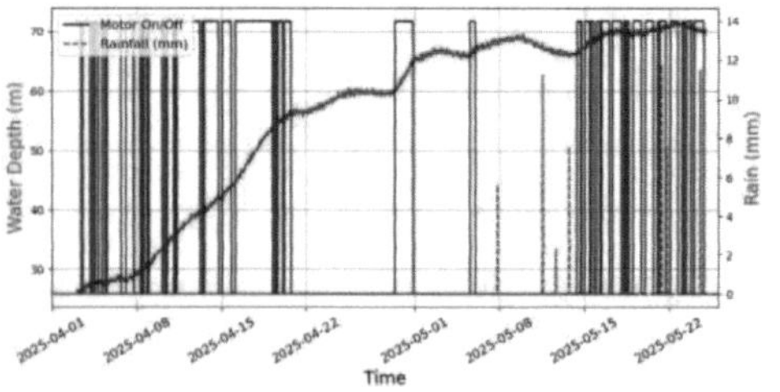

(a) Borewell A data

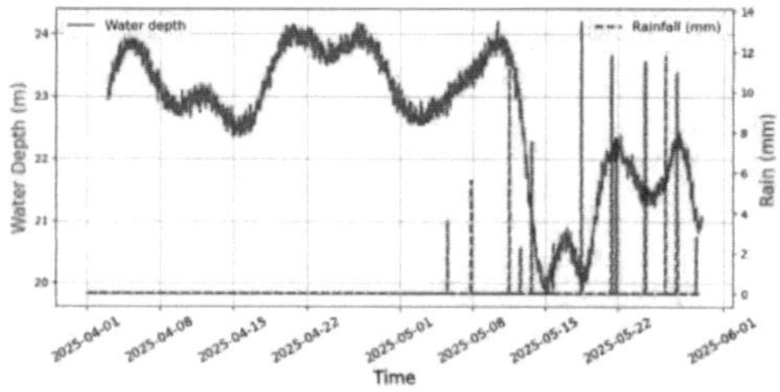

(d) Borewell D data

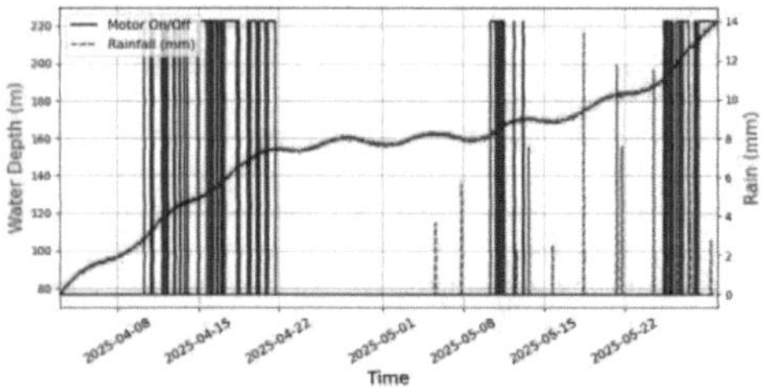

(b) Borewell B data

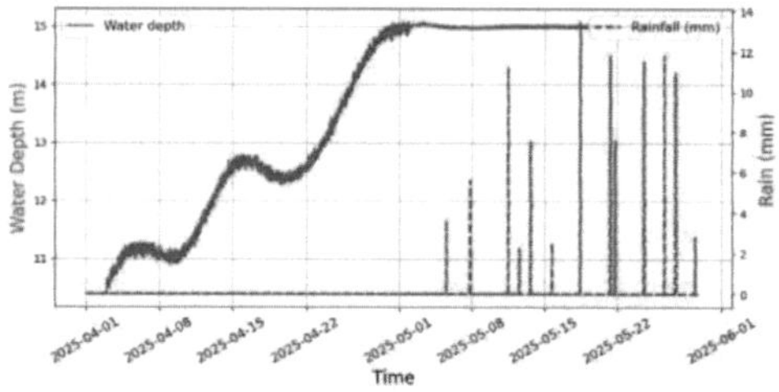

(e) Borewell E data

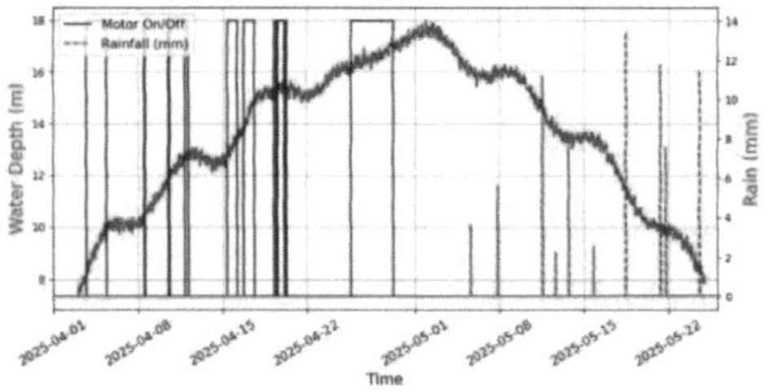

(c) Borewell C data

Fig. 4. Data from extraction and injection borewells: (1) Extraction borewells – Borewells A, B, and C (2) Injection borewells – Borewells D and E.

The trend in borewell C in Fig. 4c, again, shows an increasing trend during the month of April. A noteworthy observation here is the considerably low depth at which water is found in this borewell, as compared to the other two extraction borewells. This is due to the fact that there are multiple injection borewells and rainwater recharge pits in the vicinity of this borewell, which have led to a very healthy aquifer having a higher water level throughout the year. This borewell also recharges rapidly upon the onset of pre-monsoon and quickly reaches the pre-summer depth, once again possibly owing to very efficient recharge due to the rainwater harvesting efforts done in the vicinity.

4.1.2 Injection borewells

The injection borewells on campus, namely borewells in D and E, have the primary purpose of recharging the underground aquifers. These were also monitored during this study to see the behaviour of the water level without the effect of pumping. The results are shown in Fig. 4d and 4e

From the trend in borewell D as seen in Fig. 4d, the water level fluctuates slightly but isn't affected significantly during the summer months. With the onset of pre-monsoon, a clear decline is seen in the water depth, indicating that the surface runoff is likely to collect near this point and feed into the injection borewell. This observation is again consistent with the borewell lying in a relatively lower region in comparison to the surroundings, as can be verified from the contours in Fig. 2. The underground shallow aquifer is likely to follow these natural contours, hence resulting in prominently visible fluctuations in water depth due to rainfall at this location.

Borewell E in Fig. 4e shows an increasing trend in water depth, as was observed in the case of extraction borewells in an earlier section, even though this borewell is not being pumped. Moreover, this borewell dried up at around 15 m of depth and wasn't revived even after the onset of pre-monsoon. This observation could likely be explained by the presence of 3 new extraction borewells in the vicinity to facilitate water supply for construction activities demarcated by shaded grey regions in Fig. 2. It is possible that these new borewells and this injection borewell are connected to the same underground aquifer, and the excessive pumping, coupled with the inadequate recharge, has led to an increase in water depth and the subsequent drying up of this borewell.[1]

4.2 Water level vs precipitation

In order to quantify the relationship between rain and changes in the water depth, Pearson correlation was calculated between these two time series with varying delays in order to model the time it takes for the water to percolate and recharge the underground aquifer. The plots are shown in Fig. 5. The correlation was calculated after taking a 12-hour rolling mean of both the data frames in order to remove instantaneous fluctuations while preserving the underlying trend. It is observed from the plots of Borewells A and B that deep borewells do not show a significant positive correlation. This is attributed to the depth of the borewells increasing monotonically even after rainfall, whereas ideally, if the borewells were recharging, the correlation coefficient should have been negative. This could mean that deep borewells like borewells A and B, which do not have any rainwater channels feeding into them, are not strongly affected by rain.

---

[1] Hydrological tests conducted over multiple seasonal cycles and long-term data are required to make conclusive statements about borewell quality. The geology estimation aims to illustrate how the water level device can aid the process.

From the plots of Borewells C and D, a strong negative correlation is observed. This means that both of these borewells recharge with rain. Borewell C, which is an extraction borewell, has multiple rainwater harvesting efforts implemented around it, which are likely playing a role in the borewell showing a negative correlation. Moreover, the correlation is most negative with a delay of four days, implying that it takes about four days for the surface water to percolate and reflect as a change in the water depth in the borewell. Borewell D, which is an injection borewell, is more likely to be connected to a shallow aquifer and reflects

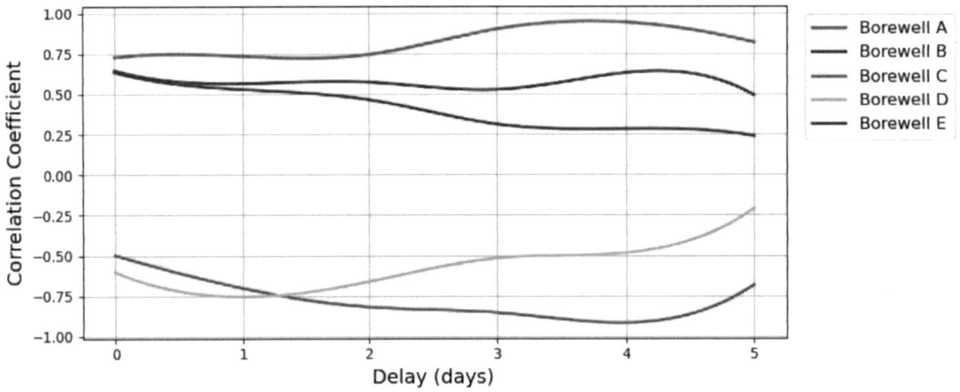

Fig.5. Correlation of water level with rain data

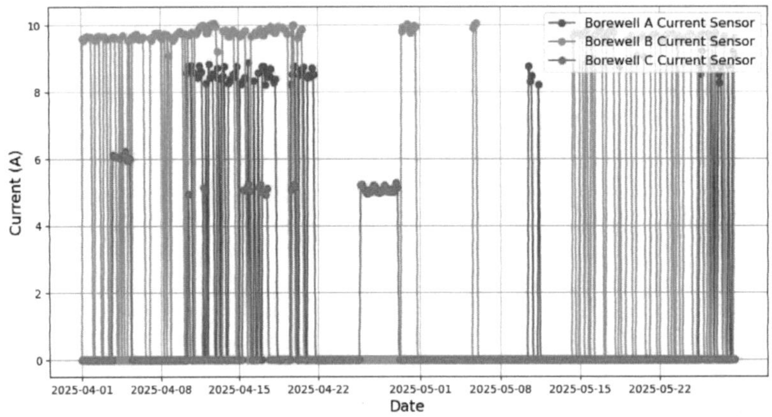

Fig.6. Motor current data

the changes in this aquifer. This can also be the reason why the most negative correlation in this case is seen at a delay of one day, as the shallow aquifer quickly recharges with surface rainfall.

The purple plot in Fig. 5 shows the correlation of injection borewell E. This borewell had become dry around the onset of pre-monsoon, and hence it should not show any correlation with rainfall, which is what we are observing here with the trend approaching 0 from the positive end. The initial positive correlation might have been caused due to the increasing depth during the summer months.

4.3 Water level vs motor current

The motor current for the extraction borewells A, B and C was also tracked. The results are plotted in Fig. 6. It is observed that the motor current is not significantly affected by the water depth. Instead, the current consumed is dependent on the rating of the motor itself, which is 6 HP, 7.5 HP and 5 HP for borewells A, B and C, respectively. Some fluctuations are observed in the current consumption but these are within the operating parameters of the motors.

5 Conclusion

This work demonstrates the potential application of IoT technologies to monitor and evaluate natural resources. Several IoT devices were deployed across five borewells to monitor the water level, motor pump usage, and rainfall. The data shows tangible observations which are in line with the geographical characteristics of the environment. Such devices can help provide insights and early warnings about changes in underground geology, as was seen in the case of borewells A and E, so that appropriate measures can be taken to safeguard these crucial resources before long-term damage. In addition to this, long-term data acquisition could potentially help clearly demarcate aquifer connectivity, which is very crucial information for planned water usage.

Another important caveat is the difference that rainwater harvesting interventions make in the overall health and recharge capacity of borewells. Such data can bring about awareness and encourage people to engage in such simple but effective techniques, which return water to the ground and aim for a net-zero where the amount of water being pumped is equal to the amount of water being returned to the underground aquifers. Future work could explore longterm trends in groundwater systems over multiple seasons. The IoT-enabled monitoring system demonstrated in this paper represents a critical step towards more informed, data-driven approaches to water resource management, environmental conservation and sustainability.

References

1. S. Vishwanath, *For sustainable borewells* (2018), accessed: 2025-06-25, https://www.thehindu.com/life-and-style/homes-and-gardens/for-sustainable-borewells/article25743716.ece
2. Arghyam and Biome Environmental Trust, *Borewells* (2018), accessed: 2025-06-25, https://www.urbanwaters.in/borewells/
3. V. Srinivasan, N.R. Lakshmikantha, G. Manjunatha, G.N. Shinde, PLOS Water 4, e0000138 (2025), accessed: 2025-06-25
4. Aderemi et al., Sustainability 14 (2022)
5. MacDonald et al., Nature Geoscience 9, 762 (2016)
6. E. Ortiz et al., Sensors 23, 9364 (2023)
7. Chan et al., Progress in Physical Geography: Earth and Environment 45, 305 (2021)
8. A.J. Calderwood, R.A. Pauloo, A.M. Yoder, G.E. Fogg, Water 12, 1066 (2020)
9. O.H. Kombo, S. Kumaran, A. Bovim, IEEE Access 9, 128417 (2021)
10. T. Bhatt, T. David, R. Yelekar, S. Chaudhari, *String Tension Based Borewell Water Level Monitoring Using IoT* (WF-IoT, 2023, 2023), pp. 1–6

11. Washington State Department of Ecology, *Integrated statewide groundwater monitoring strategy*, https://apps.ecology.wa.gov/publications/SummaryPages/1711005.html (2017, accessed 30 April 2025), publication No. 17-11-005, Chapter 9
12. *ThinkSpeak* (27 Nov, 2024), https://thingspeak.com/
13. *DFRobot 'Gravity : tipping bucket rainfall sensor'* (27 Nov, 2024), https://www.dfrobot.com/product-2689.html
14. *YHDC SCT-013-000 CT* (27 Nov, 2024), http://en.yhdc.com/product/SCT013401.html
15. Kulanuwat et al., Water 13, 1862 (2021)

CrackVision: A Bayesian CNN–Vision Transformer Bagging Ensemble for Enhancing Multiclass Concrete Crack Severity Classification

Jomer R. Mandap[1], *Daniel John Henrick D.* Sanchez[2], *Steven Mark B.* Ontanillas[3], *Neil P.* Magloyuan[4]

[1]*College of Computer and Information Science Mapúa Malayan Colleges Mindanao* Davao City, Philippines jrMandap@mcm.edu.ph
[2]*College of Computer and Information Science Mapúa Malayan Colleges Mindanao* Davao City, Philippines djhSanchez@mcm.edu.ph
[3]*College of Computer and Information Science Mapúa Malayan Colleges Mindanao* Davao City, Philippines smOntanillas@mcm.edu.ph
[4]*College of Computer and Information Science Mapúa Malayan Colleges Mindanao* Davao City, Philippines npMagloyuan@mcm.edu.ph

Abstract— This paper presents CrackVision, a bagging ensemble integrating a Bayesian Convolutional Neural Network (BCNN) and Vision Transformer (ViT) for multiclass concrete crack severity classification. Unlike binary detection systems, CrackVision categorizes cracks into four levels None, Low, Medium, High with uncertainty awareness through Monte Carlo dropout. The system was trained on 60,000 augmented crack images and evaluated against standalone models. CrackVision achieved 99.31% accuracy and F1-scores up to 99.94%, improving performance by 2.17% over BCNN and 0.25% over ViT. Confusion matrix analysis confirmed fewer misclassifications than BCNN across all severity levels. Predictive uncertainty estimates enhance reliability for safetycritical deployment. These findings highlight CrackVision's potential as a robust tool for automated infrastructure monitoring, particularly in disaster-prone regions requiring accurate crack assessment.

Keywords— Concrete Crack Severity Classification, Bayesian CNN, Vision Transformer, Ensemble Learning, Infrastructure Monitoring, Uncertainty Quantification

© The Authors, published by EDP Sciences. This is an open access article distributed under the terms of the Creative Commons Attribution License 4.0 (https://creativecommons.org/licenses/by/4.0/).

1. INTRODUCTION

1.1 Background of the Study

Concrete is the backbone of modern infrastructure, yet its durability is threatened by environmental exposure, material aging, and seismic activity. Cracks are among the earliest indicators of structural distress, and their timely detection is essential for public safety [1]. Recent earthquakes in Mindanao further underscore the need for scalable inspection solutions to enhance disaster resilience in the Philippines [1].

Traditional non-destructive testing (NDT) methods such as Ultrasonic Pulse Velocity and Ground Penetrating Radar provide reliable assessments but require specialized equipment and high operational costs, limiting large-scale deployment [2]. In contrast, AI-based image inspection enables cost-efficient and scalable monitoring using standard RGB imagery. However, most existing systems focus on binary crack detection and fail to assess severity, and many models lack uncertainty estimation, reducing reliability in safety-critical environments [3–5]. Engineering standards further emphasize the importance of severity-based maintenance prioritization.

1.2 Problem Statement

Most image-based crack detection systems remain limited to binary outputs, providing insufficient guidance for repair prioritization. These models often suffer from overfitting and degraded performance under real-world conditions, compounded by the absence of uncertainty-aware predictions [3–5]. Such limitations hinder reliable deployment in disaster-prone regions like the Philippines.

1.3 Significance of the Study

This study proposes CrackVision, a bagging ensemble that integrates a Bayesian Convolutional Neural Network (BCNN) and a Vision Transformer (ViT) for four-level crack severity classification with uncertainty awareness. By combining complementary local and global feature representations, CrackVision improves robustness, reduces variance, and delivers calibrated confidence estimates [6,7]. The framework supports data-driven infrastructure maintenance and contributes to sustainable, cost-efficient monitoring aligned with global development goals.

1.4 Objectives

This study aims to develop CrackVision, a robust ensemble framework for multiclass crack severity classification. The specific objectives are to:

a) Design and implement fine-tuned Bayesian Convolutional Neural Network (BCNN) and Vision Transformer (ViT) models for four-level crack severity classification: None, Low, Medium, High.

b) Develop CrackVision, a bagged BCNN–ViT ensemble using bootstrap aggregation to improve robustness, reduce variance, and support uncertainty-aware predictions.

c) Evaluate CrackVision's performance using accuracy, precision, recall, F1-score, confusion matrices, and expected calibration error and compare it with standalone BCNN and ViT models to validate ensemble effectiveness.

1.5 Scope and Limitations

The study focuses on RGB surface crack classification into None, Low, Medium, and High categories using augmented datasets [8]. Subsurface defects, material corrosion, and mobile deployment are excluded. Evaluation emphasizes generalization and uncertainty awareness for severity classification.

2. RELATED WORKS

Crack severity is commonly defined by crack width, with industry thresholds guiding structural intervention [9]. Yang et al. demonstrated the relationship between crack width and reinforcement corrosion [10], while Villanueva et al. applied deep learning techniques for crack severity classification [11].

Traditional inspection techniques such as Ultrasonic Pulse Velocity and Ground Penetrating Radar remain effective but are costly and unsuitable for frequent large-scale monitoring [2]. Deep learning approaches, including CNN architectures such as AlexNet, VGGNet, and ResNet, have significantly improved crack detection accuracy but are predominantly applied to binary classification tasks [3].

Ensemble learning, particularly bagging, has been shown to reduce variance and improve robustness under noisy conditions [7,12,13]. Bayesian CNNs enable uncertainty-aware predictions through Monte Carlo dropout, following the foundational work of Gal and Ghahramani [4,5]. Vision Transformers capture long-range spatial dependencies and have demonstrated superior performance in structural crack analysis [6,14 - 16].

3. METHODS

3.1 Research Design

This study used a quantitative research design to create and assess a deep learning framework for multiclass crack severity classification. The ensemble was tested for performance validation by fine-tuning Bayesian CNN and Vision Transformer models on an augmented crack dataset and combining them through bootstrap aggregation, using standard classification metrics.

3.2 Datasets

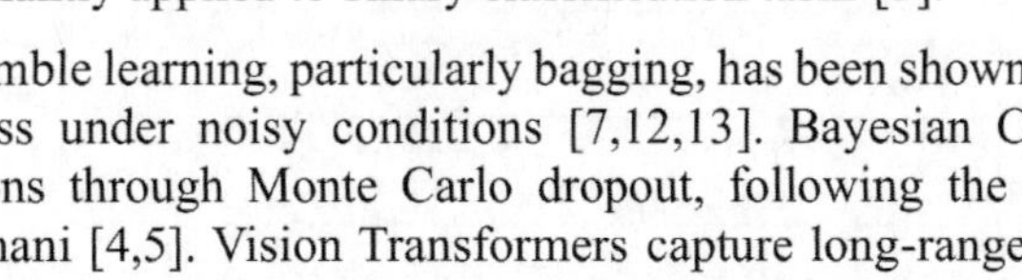

Fig.1 Sample images from Mendeley Concrete Cracking Level dataset [8]

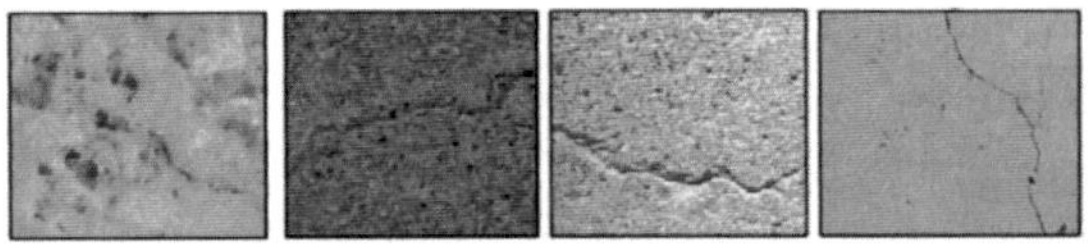

Fig.2. Sample images from USU Concrete Crack Images Dataset [15]

Two datasets were employed: the Mendeley Concrete Cracking Level dataset containing 20,000 RGB images across four crack severity levels [8], and the USU Concrete Crack Images dataset, which was used as an unseen external test set [15].

3.3 Experimental Setup

The models were trained in a controlled environment to ensure reproducibility. Table 1 summarizes the key hyperparameters, compute setup, and training configuration.

TABLE 1. EXPERIMENTAL SETUP

Component	Configuration
Environment	Google Colab Pro (GPU: NVIDIA A100)
Access Machine	MacBook Air (M1), 8 GB RAM, macOS 13.6
Framework	PyTorch 2.2, CUDA 12.1
Random Seed	42 (dataset splitting, initialization, augmentation)
Optimizer	AdamW, learning rate 1×10^{-4}, weight decay 1×10^{-4}
Batch Size	32
Epochs	15
Scheduler	Cosine annealing with warm restarts
BCNN	ResNet-18 backbone, dropout $p = 0.4$, 20 Monte Carlo forward passes
ViT	ViT-Base, patch size 16, input size 224×224
Preprocessing	Resize to 224×224, flips, rotations, color jitter
Ensemble Fusion	Weighted averaging, tuned on validation set ($\alpha = 0.5$)

3.4 Evaluation Metrics

The model's performance was evaluated using four established metrics: Accuracy, Precision, Recall, and F1-score. These metrics assess categorization efficacy across the four severity levels in relation to the ground truth labels.

Accuracy is defined as

$$Accuracy = \frac{TP + TN}{TP + TN + FP + FN} \quad (1)$$

where *TP, TN, FP, and FN* represent true positives, true negatives, false positives, and false negatives, respectively.

Precision and Recall is defined as

$$Precision = \frac{TP}{TP + FP}, \quad Recall = \frac{TP}{TP + FN} \tag{2}$$

F1-score is defined as

$$F1 = \times \frac{Precision \times Recall}{Precision + Recall} \tag{3}$$

Macro-averaging was utilized to ensure equitable treatment of all classes. Furthermore, confusion matrices were created to analyze misclassifications, and for BCNN and CrackVision, prediction variance from stochastic forward passes was documented as an indicator of uncertainty [6,7].

4. IMPLEMENTATION AND ANALYSIS

4.1 Data Preprocessing

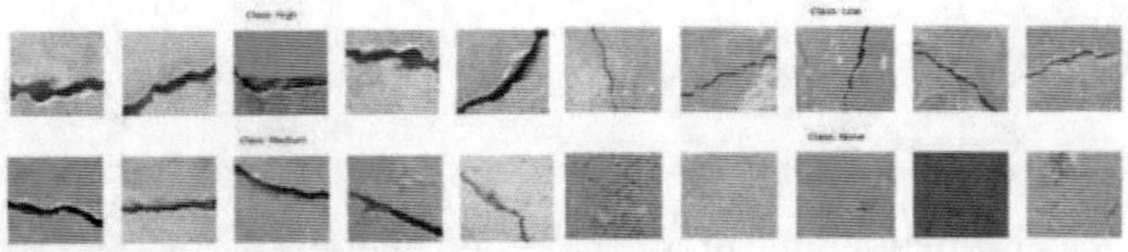

Fig.3. Preprocessed Dataset [8,15]

The primary dataset was split into 70% training, 15% validation, and 15% testing using stratified sampling. Only the training set was augmented using flips, rotations, contrast enhancement, and edge sharpening, expanding it to approximately 60,000 images. Validation and test sets remained unchanged to prevent data leakage. All images were resized to 224×224 pixels and normalized using ImageNet statistics.

4.2 System Architectures

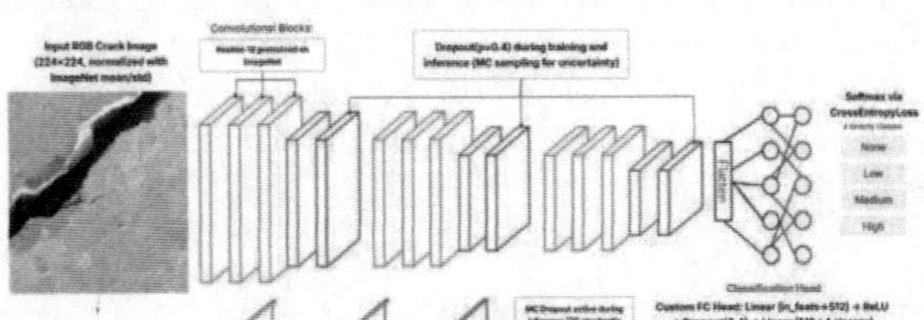

Fig. 4. Fine-Tuned BCNN System Architecture.

Fig. 4 shows the BCNN architecture used for multiclass crack severity classification. Input images (224×224) were processed using a pretrained ResNet-18 backbone, followed by a classification head with ReLU activation and dropout (p = 0.4) to produce four severity classes. Monte Carlo dropout was applied during inference to enable uncertainty-aware predictions.

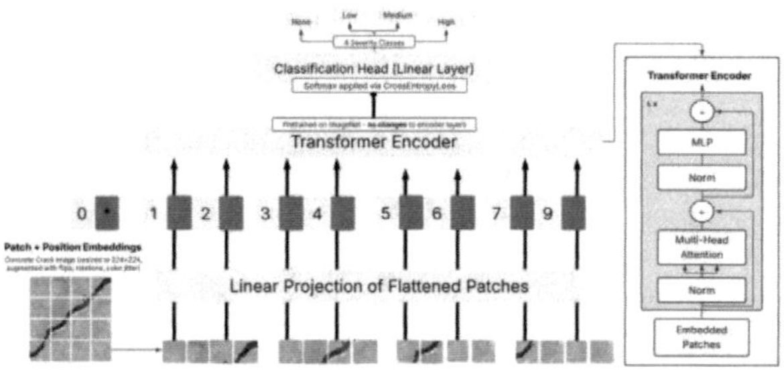

Fig. 5. Fine-Tuned ViT System Architecture.

Fig. 5 shows the Vision Transformer (ViT-Base Patch16-224) architecture fine-tuned for multiclass crack severity classification. Input images (224×224) are divided into patches with positional embeddings and processed by a pretrained Transformer encoder. A modified classification head maps the CLS token to four severity classes (None, Low, Medium, High) using softmax with cross-entropy loss.

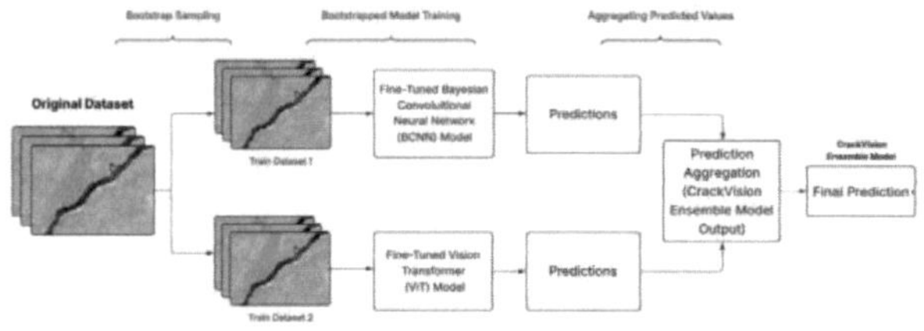

Fig. 6. CrackVision Ensemble System Architecture.

Fig. 6 shows the CrackVision ensemble architecture for multiclass crack severity classification. Bootstrap sampling is used to independently train a BCNN and a ViT, whose predicted class probabilities (None, Low, Medium, High) are combined through weighted averaging to produce the final ensemble output, improving robustness and generalization.

4.3 Base Learner Training (BCNN and ViT)

The CrackVision ensemble integrated two meticulously optimized base learners: a BCNN (ResNet-18 with dropout layers for Monte Carlo uncertainty estimates) and a ViT-Base Patch16-224 (pretrained on ImageNet and enhanced with flips, rotations, and color jitter). Both utilized cross-entropy loss, AdamW optimization, and cosine annealing scheduling. The BCNN collected local texture features with a knowledge of uncertainty, whereas the ViT modeled global structural patterns, so offering complementing capabilities for the ensemble.

4.4 CrackVision Ensemble System Training

The CrackVision ensemble was implemented using a bagging strategy that combined a Bayesian Convolutional Neural Network (BCNN) and a Vision Transformer (ViT), each fine-tuned on independently resampled bootstrap subsets of the training data to promote diversity and reduce correlated errors. The BCNN (ResNet18 backbone) incorporated Monte Carlo dropout (p=0.4) with 20 stochastic forward passes for uncertainty-aware predictions, while the ViT (Base Patch16-224) leveraged ImageNet pretraining to capture long-range spatial patterns. Their outputs were fused using a weighted averaging rule: $Pens = \alpha \cdot PBCNN + (1 - \alpha) \cdot PViT$

4.5 Model Performance

To account for statistical variability, all models were trained and evaluated across five random seeds. Results are reported as mean ± standard deviation (std).

TABLE 2. OVERALL ACCURACY COMPARISON

Model	Accuracy %
Bayesian Convulutional Neural Networks	97.14 ± 0.15
Vision Transformer (VIT)	99.06 ± 0.10
CrackVision Ensemble	99.31 ± 0.07

CrackVision achieved the highest accuracy (99.31%), outperforming BCNN (97.14%) and ViT (99.06%). The 2.17% improvement over BCNN is significant for safety-critical applications, while the 0.25% gain over ViT, though modest, remains relevant where near-perfect accuracy is required, demonstrating CrackVision's superior balance of accuracy and reliability.

TABLE 3. BCNN MULTICLASS PERFORMANCE METRICS

Class	None	Low	Medium	High
Precision	97.00% ± 0.18	98.85% ± 0.12	93.01%± 0.25	99.96%± 0.05
Recall	98.20% ±0.20	94.58% ±0.15	96.12% ± 0.21	99.70% ±0.07
F1-score	97.59% ± 0.22	96.67% ±0.22	94.54% ± 0.23	99.83% ± 0.06

The Bayesian CNN (BCNN) achieved strong overall performance with F1-scores ranging from 94.54% to 99.83%. It performed best in detecting high-severity cracks (99.83%) and non-cracked surfaces (97.59%), while showing comparatively lower performance for medium-severity cracks (94.54%).

TABLE 4. VIT MULTICLASS PERFORMANCE METRICS

Class	None	Low	Medium	High
Precision	99.36% ± 0.10	99.00% ± 0.11	99.05%± 0.12	99.00% ±0.08
Recall	99.42% ± 0.12	98.52%± 0.14	98.40%± 0.10	99.96% ±0.05
F1-score	99.40% ± 0.09	98.76%± 0.12	98.22%± 0.11	99.00% ± 0.07

The Vision Transformer (ViT) demonstrated more balanced performance than BCNN, achieving consistently high F1-scores across all classes (98.22%–99.40%). Its self-attention mechanism effectively captured spatial relationships in crack patterns, resulting in low performance variance (F1-score standard deviation: 0.58%)

TABLE 5. CRACKVISION MULTICLASS PERFORMANCE METER

Class	None	Low	Medium	High
Precision	99.30% ± 0.09	99.70% ±0.08	98.25% ±0.12	99.00% ±0.07
Recall	99.46% ± 0.08	98.78% ±0.10	99.08% ±0.11	99.00% ±0.06
F1-score	99.38% ± 0.07	99.23% ±0.09	98.67% ±0.10	99.94% ±0.18

CrackVision achieved F1-scores between **98.67% and 99.94%**, outperforming BCNN by **2.17%** and ViT by **0.25%**, while maintaining high consistency (SD = 0.54%) and strong generalization on the unseen USU dataset. The ensemble showed exceptional performance in detecting high-severity cracks (F1 = 99.94%), enabled by bootstrap resampling and an optimized fusion weight (α)..

4.6 Confusion Matrix Analysis

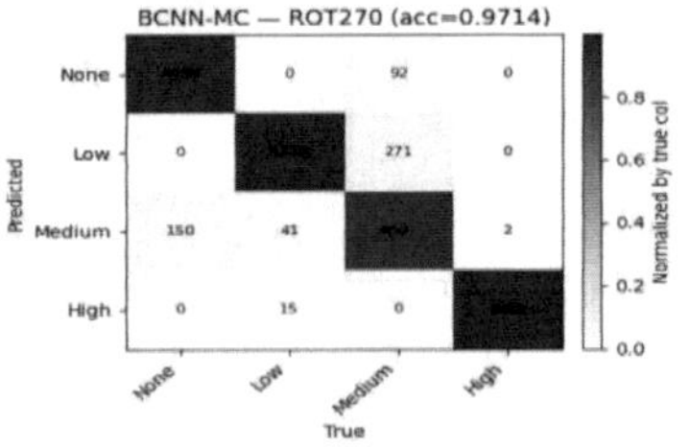

Fig. 7. BCNN Confusion Matrix.

The BCNN model showed specific confusion patterns:

1. **None class:** 4909 correctly classified, with 92 false positives misclassified as Medium
2. **Medium class:** Most challenging class with 150 samples misclassified as None and 41 as Low
3. **Cross-class confusion:** Primary confusion occurred between adjacent severity levels, indicating the model's logical progression in severity assessment

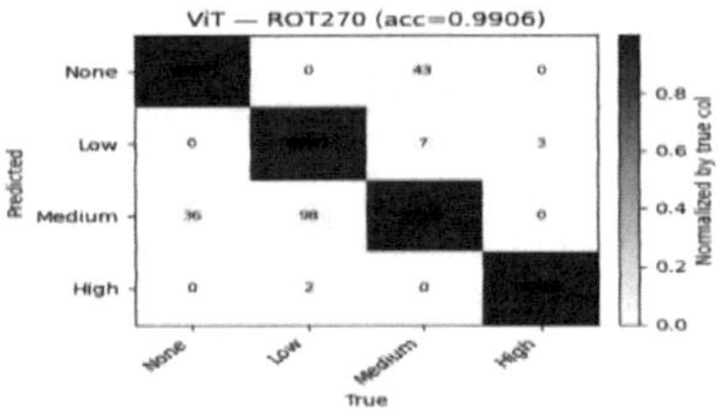

Fig. 8. ViT Confusion Matrix.

The ViT model demonstrated superior classification with minimal confusion:

1. **Balanced performance:** More uniform distribution of correct predictions across all classes
2. **Reduced medium-class confusion:** Only 36 Medium samples misclassified as None
3. **Better boundary definition:** Clearer separation between severity levels, particularly for Medium class

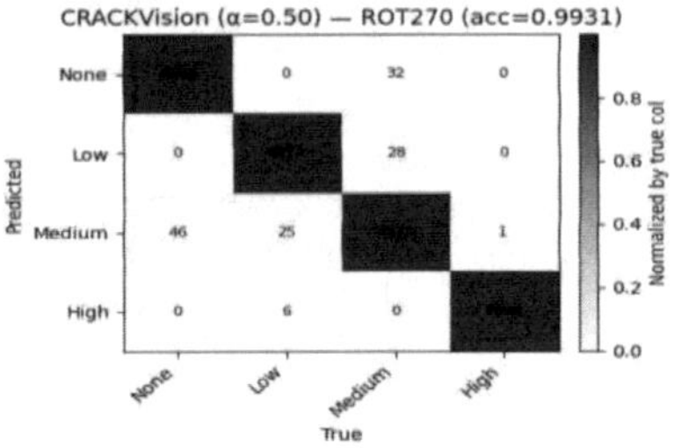

Fig.9. CrackVision Confusion Matrix

The CrackVision ensemble achieved the optimal balance:

1. **Minimal misclassifications:** Significantly reduced confusion across all class boundaries
2. **Improved medium-class detection:** Only 46 Medium samples misclassified
3. **Enhanced reliability:** Consistent performance across all severity levels

Bootstrap resampling and adaptive ensemble weighting combined BCNN's local texture sensitivity with ViT's global spatial awareness, improving overall performance. BCNN misclassified 150 medium-severity cracks (97.14% accuracy), whereas ViT reduced errors to 36 misclassifications (99.06% accuracy). CrackVision further minimized misclassifications, achieving 99.31% accuracy with balanced performance across all severity levels.

4.7 Calibration Analysis

TABLE 6. CALIBRATION RESULTS

Model	ECE %
Bayesian Convulutional Neural Networks	4.82 ± 0.12
Vision Transformer (VIT)	2.71 ± 0.09
CrackVision Ensemble	1.35 ± 0.07

Calibration was assessed using Expected Calibration Error (ECE) over five runs and reliability diagrams. BCNN showed overconfidence at intermediate probabilities, while ViT exhibited improved but slightly underconfident calibration at high confidence levels. CrackVision achieved the lowest ECE (1.35%), indicating well-calibrated and reliable predictions suitable for safety-critical applications.

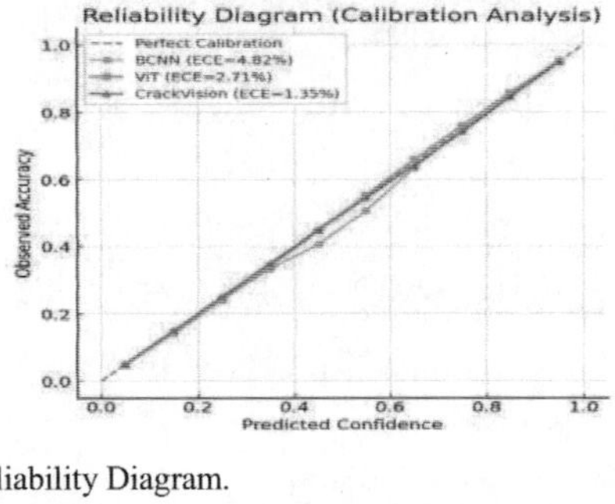

Fig.10. Reliability Diagram.

As shown in figure 10, while the BCNN curve deviated from the ideal diagonal and the ViT curve showed mild under confidence, the CrackVision curve closely followed the diagonal, demonstrating superior calibration. This indicates that the ensemble not only improved accuracy but also produced better-calibrated confidence scores, increasing its trustworthiness for real-world deployment.

5. CONCLUSION AND RECOMMENDATION

This study introduced CrackVision, a bagging ensemble combining a Bayesian CNN (BCNN) and a Vision Transformer (ViT) for multiclass crack severity classification. The model achieved 99.31% accuracy and an F1-score of 99.94%, outperforming BCNN by 2.17% and ViT by 0.25%, while providing uncertainty-aware predictions via Monte Carlo https://doi.org/50.5015/e3sconf/(0(66900100(

dropout. These results demonstrate CrackVision's potential for automated infrastructure monitoring, particularly in disaster-prone regions.

Future work will expand datasets to include diverse concrete types and conditions, explore mobile and edge deployment with real-time and offline capabilities, enhance uncertainty quantification, and extend the framework for longitudinal crack monitoring.

ACKNOWLEDGEMENT

The authors thank ACMED Builders for serving as the main consultant, Zhang Feng Qi for the Mendeley dataset, and Michael Maguire, Zhiwei Wang, and Thomas M. Johnson for the USU dataset. Special appreciation is also extended to Engineer Allan Medina for his guidance in shaping the research direction.

REFERENCES

[1] D.C.E. Llamas, D.J.A. Salcedo, C.P.R. Cabacungan, G.G.P. Salazar, D.M. Latonio, PreventionWeb (2024).

[2] F. Tosti, C. Ferrante, Surv. Geophys. **41**, 485–530 (2022).

[3] W.R. Silva, D.S. Lucena, Proc. 18th Int. Conf. Exp. Mech. (2021).

[4] Y. Gal, Z. Ghahramani, arXiv:1506.02158 (2020).

[5] A. Mesquita, A Bayesian Convolutional Neural Network Approach for Image-Based Crack Detection and a Maintenance Application, Master's Thesis, Erasmus Univ. (2020)

[6] A. Dosovitskiy et al., Proc. Int. Conf. Learn. Represent. (ICLR) (2021).

[7] Z. Fan et al., Coatings **10**, 152 (2020).

[8] Z. Qi, "Concrete cracking level," Mendeley Data, V1, (2020).

[9] M.I. Hamakareem, The Constructor (2023).

[10] X. Yang, H. Pan, X. Huang, Constr. Build. Mater. **187**, 1117–1128 (2022).

[11] A. Villanueva, E. Santos, M. Manalo, Procedia Comput. Sci. **215**, 789–796 (2022).

[12] M.A. Ganaie et al., Eng. Appl. Artif. Intell. **115**, 105151 (2022).

[13] L. Breiman, Mach. Learn. **24**, 123–140 (2021).

[14] A.M. Mayya, N.F. Alkayem, Sensors **24**, 8095 (2024).

[15] M. Maguire, Z. Wang, T.M. Johnson, USU Concrete Crack Images for Classification, Utah State Univ. Digital Commons (2020).

[16] E. Asadi Shamsabadi et al., Autom. Constr. **140**, 104316 (2022).

AUTHOR INDEX

Acidera Kendrick Cyrus 76
Adlaon Mauricio 1
Agrupis Shirley 76
Alam Andrea Marie 46
Aravind Usha K. 105
Aravindakumar C.T 105
Argotti-Valencia Eduardo 11
Bautista Hugo 11
Berroya Jean Margaret 46
Bertulfo Roselle 1
Bhatt Tanmay 152
Calderon Aldrin 76
Chaudhari Sachin 152
Cuadrado Jerry 1
Degamon Louella 1
Dela Cerna Monalee 1
Dewati Regyna Elrica 96
Dwi Novemyanto Alfin 36
Ecle Kier 1
Eiangmee Orranat 60
Erasga Eden S. 88
Eviota Manny 1
Fan Minjia 115
Ghaida Azka 96
Goyal Prachi 96
Hieu Huynh Thuan 27
Ho Huynh Kim 27
Hugo Rosalia 1
Hutangkoon Thunyanat 60
Jiménez-Pozo Patricio 11
Kunyuan Jumpon 60
Kusuma Ririn Radiawati 96
Lal Naveen S. 105
Llames Angela Mae A. 88
Luong Nguyen Duc 27
Magloyuan Neil P. 163
Maikaew Jatuphat 60
Mandap Jomer R. 163
Manzano Loreli Faye 76
Mateo Nathaniel Ericson 76
Meilani Retno 36, 134, 143
Mosa Carmelin 1
Muharrami Revia 96
Ningsih Almira Solehah 96
Novemyanto Alfin Dwi 134
Nur Rismawati 36, 134
Odtojan Medielyn 1
Ontanillas Steven Mark B. 163
Panaligan Tristan Roy 46
Permadi Didin Agustian 96
Pragada Tarun 152
Rajurkar Prashant 17
Ramesh Kalpana 152
Reddy Jagan M. 152
Sanchez Daniel John Henrick D. 163
Sangab Kim Z. 88
Sokjabok Siwakon 60
Srisang Naruebodee 60
Srisang Siriwan 60
Suasnavas Darwin 11
Torres Princess Lynelle Carysse 46
Xia Yu 52
Xu Linglong 115
Y Nguyen Nhu 27
Yang Jianing 115
Yao Shun 52
Yap Angelica Nala T. 88
Yu Lijiao 52
Zhu Ruiliang 52